Influence of fibre orientation on fatigue of short glassfibre reinforced Polyamide

Stellingen behorende bij het proefschrift

Influence of fibre orientation on fatigue of short glassfibre reinforced Polyamide

Jaap Horst

Oktober 1997

1. De uitspraak dat in vezelversterkte materialen de scheurinitiatie in vermoeiing niet belangrijk is, omdat er aan de vezel-uiteinden altijd wel een scheurtje aanwezig zou zijn, en daarom de levensduur door scheurgroei bepaald zou worden, is onjuist. Bij deze kleine scheurtjes is de breukmechanica niet geldig, resultaten uit scheurgroeimetingen zijn dus zeker niet toepasbaar in dit geval.
    Dibenedetto A.T., Salee G. in: *Pol. Eng. Sci. 19, 1979, pp. 512-518.*

2. De laag waar eerst schade optreedt (al dan niet in vermoeiing) is niet alleen afhankelijk van de dikte van de lagen, maar ook van de volume-fractie vezels, en vindt zeker niet altijd plaats in de laag waar de vezels in belastings-richting liggen.
    Dit proefschrift, Hitchen S.A., Ogin S.l. in: *Comp. Sci. Tech. 47, 1993, pp. 83-89.*

3. Ook bij vezels evenwijdig aan de belastingsrichting kan een trekspanning op de vezel-matrix interface ontstaan, als gevolg van de dwarscontractie van de matrix bij (plastische) deformatie.

4. Het feit dat de levensduur correleert met de toename van verlenging van het proefstuk tijdens vermoeiing, geeft aan dat de vezeluiteinden veel ernstiger spanningsconcentraties veroorzaken dan eventuele beschadigingen, insluitsels etc.

5. De voor karakteristatie van de vezeloriëntatie veel gebruikte "in plane orientation" is een oversimplificatie van de werkelijkheid, zeker bij grotere diktes, zoals in het gemeten spuitgietstuk van 5.75 mm dikte.
    Dit proefschrift, Toll S., Anderson in: *Pol. Comp. 14, 1993, pp. 116-125.*

6. Het verband tussen vermoeiingslevensduur en kruipsnelheid kan een versnelling te weeg brengen bij het bepalen van vermoeiingsgedrag van producten.
    Dit proefschrift, Casado et al. in: *Poceedings of the 1st Int. Conference on Fatigue of Composites* (Ed. S. Degallaix at al.), *1997, pp. 456-461.*

7. Bij de evaluatie van het milieu-effect van automobielen, wordt veel meer gelet op de emissies tijdens gebruik dan tijdens de productie. Het is maar sterk de vraag of het inruilen van een oudere auto voor een nieuwe een positieve invloed heeft.

8. Zonder zogenaamd ontwikkelingshulp en missionariswerk waren vele zg. derde wereld landen veel beter af geweest dan nu het geval is.

9. De term 'derde wereld landen' duidt op een ongelofelijke arrogantie van ons, de 'eerste' wereld landen.

10. De religie zou ineens weer een stuk populairder worden als atheïsten geen vrij zouden hebben op de christelijke feestdagen.

11. De door de wetenschap bereikte gemiddelde verlenging van de duur van het leven van de mens, legt slechts een extra nadruk op de toenemende leegte daarvan.

12. De staat is het enige bedrijf dat als oplossing voor een tekort, niet de kosten hoeft te verlagen, maar ertoe over kan gaan de prijzen te verhogen terwijl het produkt gelijk blijft.

13. Daar veel besluitvorming verschoven is van het parlement naar het centraal planbureau en verschillende onderzoekscommissies, is het voor het behoud van de democratie noodzakelijk dat ook deze organen democratisch gekozen gaan worden.

14. Wachttijden op stations, en daarmee vertragingen in het treinverkeer, zouden sterk afnemen wanneer het instappen aan het ene eind, en het uitstappen aan het andere eind van een wagon zou gebeuren, zoals dat bij bussen en trams al heel lang gebruikelijk is.

Influence of fibre orientation on fatigue of short glassfibre reinforced Polyamide

PROEFSCHRIFT

ter verkrijging van de graad van doctor
aan de Technische Universiteit Delft,
op gezag van de Rector Magnificus Prof.dr.ir. J. Blaauwendraad,
In het openbaar te verdedigen ten overstaan van een commissie,
door het College van Dekanen aangewezen,
op dinsdag 28 oktober 1997 te 13.30 uur

door

Jakob Jan HORST
Ingenieur Werktuigbouwkunde

geboren te Winschoten

Dit proefschrift is goedgekeurd door de promotor:

Prof. Ir. J.L. Spoormaker

Samenstelling promotiecommissie:

Rector Magnificus	Technische Universiteit Delft, voorzitter
Prof.Ir. J.L. Spoormaker	Technische Universiteit Delft, promotor
Prof.Dr.Ir. A. Bakker	Technische Universiteit Delft
Prof.Dr.Ir. R. Marissen	Technische Universiteit Delft
Dr.Ir. E.A.A. van Hartingsveldt	DSM Research
Prof.Dr.Ir. A. Posthuma de Boer	Technische Universiteit Delft
Prof.Dr. P.C. Powell	Universiteit Twente
Prof.Ir. L.B. Vogelesang	Technische Universiteit Delft

Published and distributed by Delft University Press, Mekelweg 4, 2628 CD Delft, the Netherlands
Telephone: +31 15 278 3254
Telefax: +31 15 278 1661
E-mail: DUP@DUP.TUDelft.NL

ISBN 90-407-1532-7 / CIP

Copyright © 1997 J.J. Horst

All rights reserved.
No part of the material protected by this copyright notice may be reproduced or utilized in any form or by any means, electronic or mechanical, including photocopying, recording or by any information and storage and retrieval system, without permission from the publisher: Delft University Press

Printed in the Netherlands.

Aan Erika

CONTENTS

Page

List of symbols		xi
List of abbreviations		xiii
1	**Introduction**	1
1.1	Plastics Composites Materials	1
1.2	Research objective	3
1.3	Outline of the Thesis	3
2	**Orientation**	5
2.1	Introduction	5
2.2	Orientation mechanisms	6
2.3	Orientation description methods	11
2.4	Orientation measurement	13
2.5	Orientation prediction	13
2.6	Conclusions	15
3	**Mechanical behaviour of composites**	17
3.1	Introduction	17
3.2	Theoretical evaluation of Modulus with fibre orientation	18
3.2.1	Theoretical determination of Elastic Modulus in fibre direction	18
3.2.2	Calculation of effective Modulus	19
3.2.3	Stress distribution for the non-elastic case	20
3.2.4	Elastic Modulus in not longitudinal direction	21
3.3	Theoretical prediction of strength	22
3.3.1	Strength of unidirectional, short fibre composites	22
3.3.2	Strength in non-fibre direction	24
3.4	Comparison of properties with predictions	25
3.4.1	Stiffness	25
3.4.2	Strength	28
3.5	Mechanical properties of specimens used in this research	29
3.5.1	Influence of orientation	30
3.5.2	Influence of humidity	32
3.5.3	Influence of strain rate	33
3.6	Conclusions	34
4	**Fatigue Lifetime measurements**	37
4.1	Introduction	37

4.2	Experimental procedure and materials	37
4.3	Results	38
4.3.1	S-N curves	38
4.3.2	Effect of frequency	39
4.3.3	Creep Speed method	41
4.3.4	Modulus and Energy dissipation	44
4.4	Effect of orientation, normalised S-N curves	45
4.5	Effect of humidity	48
4.6	Effect of bonding and fibre length	50
4.7	Conclusions	52
5	**Fatigue Crack propagation experiments**	55
5.1	Introduction	55
5.2	Experimental procedure and materials	55
5.3	Results and discussion	57
5.4	Conclusions	61
6	**Microfoil Tensile Tests for obtaining strength profiles**	63
6.1	Introduction	63
6.2	Experimental procedure and materials	63
6.3	Microfoil Tensile Test profiles	65
6.3.1	Form of the MFTT profile, influence of matrix	69
6.3.2	Influence of fibre orientation on MFTT profiles	70
6.3.2.1	Profiles of thin (2 mm) specimens	70
6.3.2.2	Profiles of thick (5.75 mm) specimens	73
6.4	Influence of fatigue on MFTT profiles	77
6.4.1	Effect of number of cycles and load level	78
6.4.2	Effect of type of specimen	82
6.4.2.1	Profiles of thin (2 mm) specimens	83
6.4.2.2	Profiles of thick (5.75 mm) specimens	87
6.4.2.3	Synopsis	89
6.5	Conclusions	91
7	**Fractography**	93
7.1	Introduction	93
7.1.1	Fractography of SFRTP in literature	93
7.2	Experimental	94
7.3	Fracture surfaces of conditioned specimens	97
7.3.1	Tensile tested specimens	97
7.3.2	Crack growth experiments	98
7.3.3	Lifetime measurements	101
7.3.3.1	Comparison of fracture surfaces from fatigue and tensile experiments	104
7.3.3.2	Broken Fibres initiating matrix failure	105

7.4	Influence of conditioning	107
7.4.1	Appearance of fracture surface	107
7.4.2	Relation between microductile area of fracture surface and load level	111
7.4.2.1	Modelling	112
7.4.2.1	comparison of results with modelling	113
7.5	Influence of improved fibre coating	114
7.6	Cryogenically broken pre-fatigued specimens	117
7.6.1	Transverse cryogenic fracture	118
7.6.2	Longitudinal cryogenic fracture	122
7.7	Conclusions	127
8	**Failure Mechanism**	129
8.1	Introduction	129
8.2	Explanation of failure mechanism	130
8.2.1	Conditioned material	130
8.2.2	Saturated material	132
8.2.3	Dry as moulded material	133
8.2.4	Material with improved bonding	133
8.3	Modelling of tensile debonding	133
8.4	Implications of the failure mechanism	137
8.5	Modelling of fatigue behaviour using Master Curves	139
8.5	Conclusions	143
9	**Practical applications of the research**	145
9.1	Introduction	145
9.2	Design rules for fatigue	145
9.3	Mould design	147
9.4	Influence of processing conditions	152
9.5	Fatigue Data	152
9.6	Conclusions	153

Summary	155
Samenvatting	159
References	163
Appendix I Injection moulding conditions	173
Appendix II Fatigue test sheet	174
Acknowledgements	175
About the author	177

List of Symbols

Symbol		Unit
A	constant in the Paris Law	mm/cycle
A'	constant in modified Paris Law	mm/cycle
A_f	cross section of fibre	mm^2
a	crack length, half the crack length for a centre crack	mm
$a_{ij(kl)}$	elements of the orientation tensor	-
a_n	fraction of fibres with orientation θ_n	-
da/dn	crack growth rate	mm/cycle
E_{90}, E_T	transverse elastic modulus	MPa
E_f	elastic modulus of fibre	MPa
E_L	composite elastic modulus in fibre direction	MPa
E_m	elastic modulus of matrix	MPa
E_n	elastic modulus of layer n	MPa
E_T	composite elastic modulus perpendicular to fibre direction	MPa
E_θ	composite elastic modulus at angle θ to fibre direction	MPa
F	force in the fibre	N
G	shear modulus	MPa
G_m	shear modulus of matrix	MPa
G	(Ch. 5) strain energy	MPa·m
H	constant	MPa
f_p	orientation parameter (planar orientation)	-
g_p	orientation parameter (planar orientation)	-
l	fibre length	µm
l_c	critical fibre length	mm
l_d	distance to debonded area	µm
l_m	modelled length in FEM	µm
M	microstructural efficiency factor	-
m	exponent in the Paris Law	-
m'	exponent in modified Paris Law	-
N	number of cycles to failure	-
n	number of cycles	-
$\underline{p}$	unit orientation vector	-
R	minimum to maximum load ratio	-
R	(Ch. 8) half the distance between fibres	µm
R_v	distance between fibres (heart - to - heart)	mm
r	fibre radius	µm
S	stress amplitude	MPa
T_g	glass transition temperature	°C
t	total thickness laminate-composite	mm
t_n	thickness laminate with orientation θ_n	mm
u	elastic displacement of fibre	mm
V_c	cyclic creep speed in secondary creep	mm/cycle
v	elastic displacement of matrix	mm

v_f	fibre volume fraction	-
W	width of the specimen, half the width for a centre cracked specimen	mm
x	coordinate along fibre	mm
Y	geometry factor	-
β	fibre effectiveness factor	-
γ	shear strain	-
ΔG	strain energy release rate	MPa·m
ΔK	stress intensity difference	MPa·√m
ϵ	strain in matrix	-
η	efficiency factor $\eta = \eta_o \eta_l$	-
η_o	orientation efficiency factor	-
η_l	length efficiency factor	-
θ	angle between fibre direction and load direction	rad
θ, ϕ	orientation of fibre in the coordinate system (Ch2)	rad
ν	poisson's ratio	-
σ_c	residual strength of a specimen with a crack	MPa
σ_f	stress in fibre	MPa
σ_m	stress in matrix	MPa
σ_m'	stress in matrix at fracture strain of fibre	MPa
σ_{max}	maximum stress in fatigue experiment	MPa
σ_{min}	minimum stress in fatigue experiment	MPa
σ_{uc}, σ_L	composite strength, fibre direction	MPa
σ_{uf}	fibre strength	MPa
σ_{ut}, σ_T	composite strength, transverse direction	MPa
$\sigma_{u\theta}$	composite strength at an angle θ between fibre and load direction	MPa
$\bar{\sigma}_f$	average stress in fibre	MPa
$\bar{\sigma}_{uf}$	average stress in fibre at fibre fracture	MPa
τ	shear stress	MPa
τ_y	shear yield stress	MPa
τ_u	shear strength of interface or matrix	MPa
τ_{uc}, τ	in plane shear strength of composite	MPa
ψ	distribution function of fibre orientation	-

subscripts:
c   of the composite
f   of the fibre
m   of the matrix
u   maximum (ultimate)

0   in fibre direction
90   perpendicular to fibres

Thus: σ_{um} is the maximum stress in the matrix

List of Abbreviations

GFPA	Glassfibre reinforced PolyAmide
GFPE	Glassfibre reinforced PolyEthylene
GFPP	Glassfibre reinforced PolyPropylene
MFD	Mould Flow Direction
MFTT	Microfoil tensile test
FCP	Fatigue Crack Propagation
FEM	Finite Element Method
PP	PolyPropylene
PA	PolyAmide
PBT	PolyButeneTerephtalate
SFRTP	Short fibre reinforced thermoplastic
UTS	Ultimate tensile strength

1. Introduction

1.1 Plastics Composites Materials

A considerable disadvantage of unreinforced plastics is their low stiffness and strength. The use of plastics has therefore been confined to domestic articles, not being loaded by severe mechanical forces. The addition of particles, however, greatly increases the stiffness of plastics and especially the addition of reinforcing *fibres* (carbon, aramid and glass) increases the strength as well, at least in the fibre direction. Fibre reinforced plastics are used for highly loaded parts, like for example under-bonnet applications in automobiles. The length of the reinforcing fibres dominates the properties, longer fibres resulting in higher strength and stiffness.

Many materials have been developed where the fibres are used in a continuous manner, thus taking maximum advantage of their reinforcing effect. Manufacturing is limited to shell-like structures, and can be done by filament winding, hand lay-up and sheet forming processes. However, in injection moulding short fibres have to be used. This mix of thermoplastic matrix and short fibres is known as Short Fibre Reinforced ThermoPlastics (SFRTP's). Average length of the fibres in the final product will be in the range of 0.2 - 0.3 mm. Recent developments can increase this value up to a maximum of 10 mm. By using the pultrusion technique to compound (mix) fibres and polymer, fibre fracture in the compounding step is avoided. Consequently injection moulding parameters and mould design have to be optimised for minimum fibre breakage during processing.

While Carbon and Aramid are known as fibres for high performance materials as used in aviation industry, the glassfibre is very attractive because of its low cost, and is widely used in reinforcing engineering plastics like PP, Nylon (PA6 and PA6.6) and PBT. Generally the per-kilogram price of glassfibres is comparable to that of the plastic it is used to reinforce. Thus the price of the reinforced grade is mainly increased by the compounding step, needed to pre-mix the fibres and the polymer.

A characteristic of SFRTP's is their high degree of anisotropy, caused by fibre orientation, see Fig. 1.1. Even in the case of a simple injection moulded SFRTP plate a layered structure exists: Skin, shell and core layers can be distinguished, with orientations usually random, aligned with Mould Flow Direction (MFD) and perpendicular to MFD respectively. Orientation in these layers as well as the thicknesses of the layers vary from location to location in the plate. Therefore the material properties vary throughout the plate (Fig. 1.1). For example the tensile strength of specimens cut from a plate can vary between 100 and 160 MPa. This depends on the location from where the specimens are cut, and the direction of the axis of the specimens relative to the Mould Flow Direction (MFD). Modulus of Elasticity of the specimens varies by approximately the same degree as the strength.

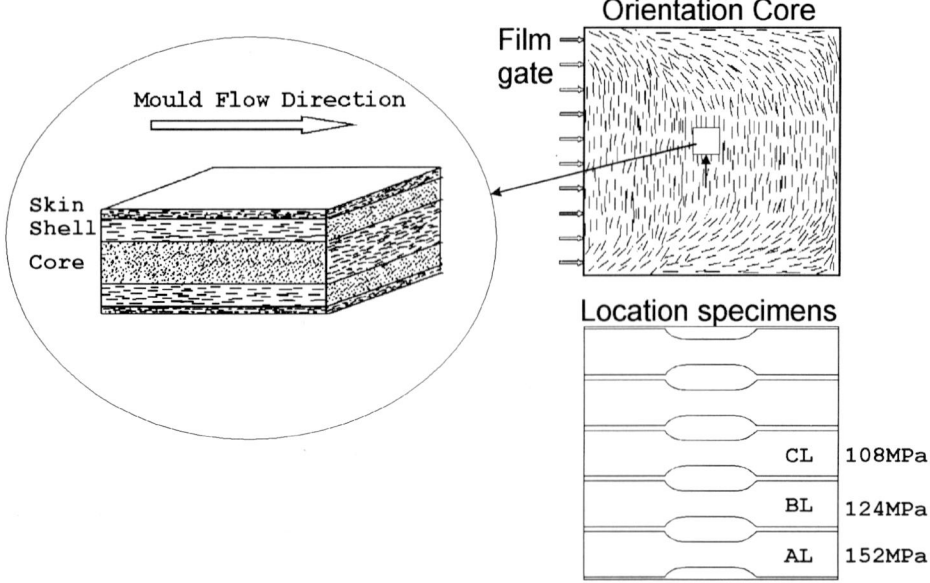

Figure 1.1   Fibre orientation as seen in the injection mouldings used in this research. Even a simple geometry like a square plate possesses a complex orientation, leading to anisotropy and inhomogeneity of properties. The strength for the shown specimens is given, in the case of a conditioned plate.

Fatigue of SFRTP's has been studied extensively, both fatigue crack growth [Karbhari,90, Karger-Kocsis,88, Mandell,83, Lang,81, Dibenedetto,79] and lifetime measurements on unnotched specimens [Adkins,88, Mandell,83, Mandell,80]. The problem with the majority of these investigations, when trying to use the results in engineering applications, is the way the orientation has been treated. Many of the research has been done using injection moulded specimens, thus ignoring fibre orientation effects as all specimens have the same fibre orientation. A number of investigators have taken the fibre orientation inside their specimens into account [Wyzgoski,94, Wyzgoski,92, Darlington,91, Voss,88, Lang,87, Friedrich,86, Carling,85, Lang,83], however in most cases this is not detailed enough to be able to predict the fatigue behaviour of real structures. In other cases only fatigue crack propagation has been studied [Friedrich,86], ignoring the crack initiation stage. Friedrich et al used a fibre effectiveness approach, in which fibre length, fibre content and fibre orientation was accounted for. Many investigators treated the orientation in a simplified way; only Longitudinal and Transverse (to Mould Flow Direction) specimens were distinguished. To be able to predict the fatigue performance of injection moulded products, first of all the knowledge of the effect of orientation on fatigue behaviour is to be extended.

1.2 Research objective

The goal of the research presented here, is to provide design rules for assessing the fatigue behaviour of parts, injection moulded from fibre reinforced Polyamide 6 in particular. Of course it would be desirable if these rules could be extended for reinforced plastics with a different matrix polymer as well.

The fatigue behaviour of parts, with varying fibre orientation as brought about by the injection moulding process, is not easily predicted. The material properties are anisotropic, and inhomogeneous: Because the fibre orientation varies throughout the part, the properties vary within one product. Prediction of properties thus requires prediction of fibre orientation. A substantial increase in the confidence in the prediction of fibre orientation using Finite Element Methods is expected in the near future. Orientation prediction however was *not* part of this research, it will be discussed however, in Chapter 2.

Besides a well-established method to obtain the fatigue behaviour of a part, if the orientation is known (the main research objective). A second objective is to give general rules for mould design, as to avoid fibre orientations known to give low fatigue performance, and get favourable fibre orientation for good fatigue behaviour.

Finally the failure mechanism (or mechanisms), in fatigue should be understood. Understanding of the failure mechanism will inform us about the general applicability of the research, and thus enable us to transfer the results found in this research to fatigue behaviour of other SFRTP's, or material tested under different conditions.

1.3 Outline of the Thesis

In Chapter 2 the fibre orientation will be discussed, from the way it is formed during mould filling, to the recent developments in the prediction of the fibre orientation. The influence this orientation has on various mechanical properties will be discussed in Chapter 3, including methods to predict these properties, when the fibre orientation is known. Results of the investigation of fatigue behaviour, as it depends on fibre orientation or properties derived from fibre orientation, will be presented in Chapters 4 and 5, discussing the fatigue behaviour in lifetime measurements respectively crack growth experiments. All the experiments have been executed in *tension*, not in bending, because in tension a simple stress distribution and uniform strain through the thickness of the specimens is obtained.

The fatigue results of Chapters 4 and 5 give information which is useful to the designer. However, the scientist is not satisfied unless an explanation for the results is given. The failure mechanisms that occur in fatigue can give this explanation. Various methods can be used, which can be divided into two categories: Observations during the fatigue process, and observations after fatigue fracture has occurred.

To the first category belong measurements of mechanical properties during fatigue, like measurements of creep speed, elastic Modulus and energy dissipation, as discussed in Chapter

4. Characteristic of these is that the measurements do not interfere with the fatigue process. The fatigue experiment can also be stopped to obtain information on changes in the specimen due to the fatigue process. The specimen can for example be broken to reveal the microscopic structure inside, see Chapter 7. Also changes in properties of the specimen can be measured in a destructive manner [Kalinka,90]. Generally if the fatigue experiment is stopped, the number of cycles the specimen would have resisted under fatigue loading is not known. This number can vary considerably, reason why a good prediction of this value is needed. The creep speed method presented in Chapter 4 can be used to make a prediction of the lifetime of a specimen under fatigue, during the fatigue experiment.

The main method in the second category is fractography, the microscopic observation of the fracture surfaces, see Chapter 7. Drawback of this method is that the features visible on the fracture surface are the result of the entire failure mechanism. The events occurring during the failure thus can be obscured by changes that occur later in the fatigue process.

In Chapter 6 an innovative method in the first category will be discussed. Micro Foil Tensile Tests (MFTT) are used to obtain strength profiles after fatigueing. Thus the decrease in strength and the change in fracture strain is measured, *in the individual layers* (core, shell skin). The result is an impression of the layers in the thickness where fatigue damage occurs first.

In Chapter 8 the Failure Mechanism and modelling of the fatigue behaviour and fibre - matrix debonding will be treated. In the last Chapter (9) the implications of the research for practical design purposes will be given.

2. Orientation

2.1 Introduction

In injection moulding of short glassfibre reinforced thermoplastics no direct influence on fibre orientation during the production is possible, as is the case with continuously reinforced glassfibre composites. Early articles on fatigue of SFRTP totally ignore fibre orientation [Mandell,80], while later articles do recognise that a certain fibre orientation exists, but no connection to the behaviour of glassfibre reinforced thermoplastics is made. Often the same type of specimens were used as in research on not-reinforced polymers. As these are normally injection moulded dog-bone type specimens, with high alignment of fibres, upperbound values for the material properties are obtained. The variation of properties with orientation was not considered.
The fibre orientation is formed during mould filling and packing, causing the fibres to orientate according to the flow of polymer melt in the mould and runner system. The fibres generally possess an average direction, causing strong anisotropy of physical and mechanical properties of the injection moulded short fibre reinforced products.

If the fibre orientation would be known exactly, theoretically it would be possible to calculate the material properties like strength and stiffness as well. Advances are made in realising this goal, see Chapter 3. In the ideal case the fibre orientation and mechanical properties resulting from the mould design should be evaluated before the mould is manufactured. Various commercial programs are on the market to evaluate mould design with respect to filling, cooling, warpage, void formation etc for non-reinforced plastics. The same would be desirable for fibre reinforced plastics. Commercial programs like C-mold or Mold-flow already include a fibre-orientation module. However predictions are not as accurate as desired. A great deal of research in this field is currently being done, as will be discussed in paragraph 5 of this chapter.

An effect that is generally not taken into account when studying the orientation of fibre reinforced plastics, is the molecular orientation that can be present in the polymer matrix. Molecular orientation in unreinforced polymers is a well-known effect, known to influence the mechanical properties of parts injection moulded of both amorphous and semi-crystalline polymers [McCrum,88]. For reinforced polymers however I know of only two studies where this effect has been studied [Yu,94, Folkes,80]. Molecular orientation was proven using birefringence measurements, in both GFPP and GFPE. Largely the molecular orientation follows the orientation of the fibres, with the difference that molecular orientation in molten polymer relaxes quickly, while the fibre orientation will not relax. Part of the molecular orientation is caused not by the polymer flow, but by the growth of crystals initiating at the fibres [Devaux,90]. Thus part of the molecular orientation is a reflection of the fibre orientation. The magnitude of the molecular orientation is comparable to that observed in unfilled polymers. Few research has been done on the effect of this molecular orientation on the mechanical properties of injection moulded products, generally it is considered to be relatively low. Through thickness measurements by O'Donnell and White in 1994 show that it can lead to a considerable increase in Elastic Modulus at the surface, and a decrease in Modulus directly below the surface. Computations of stiffness from fibre orientation showed that the variations in matrix stiffness must be included in the calculations to get a good comparison with actual measurements. For the matrix properties measurements on pure PA were used. The effect of matrix molecular orientation and/or variations in crystallinity will not be considered in this thesis.

In this chapter first a short overview will be given of the mechanisms in mould filling and melt flow that cause the fibre orientation. The basic principles of fibre orientation are explained, consequently the events during mould filling are discussed. Fibre orientation is not a simple quantity, nor can the fibre orientation for each individual fibre be given. Therefore a number of methods to mathematically describe fibre orientation have been developed. Knowledge of the orientation is crucial for prediction of the material properties in an actual product. The fibre orientation can be either measured in the final product, or predicted using the FEM methods mentioned before in the design stage. A short overview will be given in the last paragraph of this chapter.

2.2 Orientation mechanisms

The orientation of fibres in a viscous flow is caused basically by differences in velocity of both fibre ends. The velocity differences of the fibre are considered to be caused by differences in polymer flow velocity. Especially in thermoplastics with high melt viscosity, the fibre velocity will be equal to the polymer flow velocity. Thus in a homogeneous flow no change in orientation

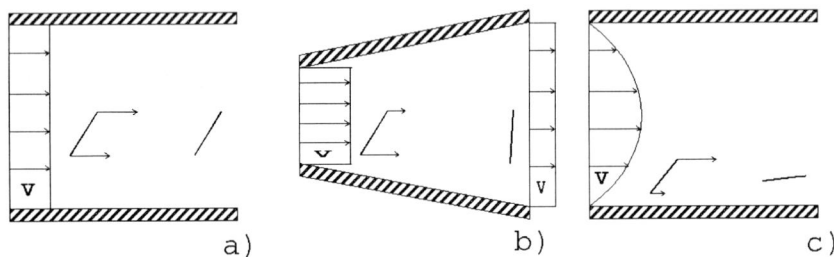

Figure 2.1   Orientation effects: a) Homogeneous flow, b) Diverging flow, c) shear flow

of fibres will take place (Figure 2.1a). When a flow is not homogeneous, two basic effects can be distinguished: The variation in velocity can be in flow direction, or perpendicular to flow direction. These are known as divergent/convergent flow respectively shear flow (Figure 2.1b and c). Diverging and converging flow occur when the cross section of the mould cavity increases respectively decreases in flow direction. Shear flow occurs also in cavities with constant cross section, and is caused by the adhesion between polymer melt and mould wall. This causes the polymer velocity to be zero at the mould wall, leading to a velocity profile over the cross section. In a Newtonian fluid this would be a parabolic profile. The fibre orientation in dilute suspensions in a shear flow is not in flow direction. According to the Jeffery equation [Jeffery,22], the fibres should continue rotating, however the fibres spend most of the time orientated in flow direction, before making a turn again. In concentrated suspensions (as used in engineering plastics) the fibres are not supposed to make these turns [Folgar,84]. However Folgar et al. do observe full fibre rotations (in shear) in concentrated suspensions as well.

The resulting fibre orientation due to each type of flow is:
In a diverging flow the fibres will orientate perpendicular to flow direction, while in converging flow and shear the fibres will orientate parallel to flow direction.
The converging flow was reported to be 30 times more effective in aligning fibres in flow direction than shear flow [Baraldi,92].

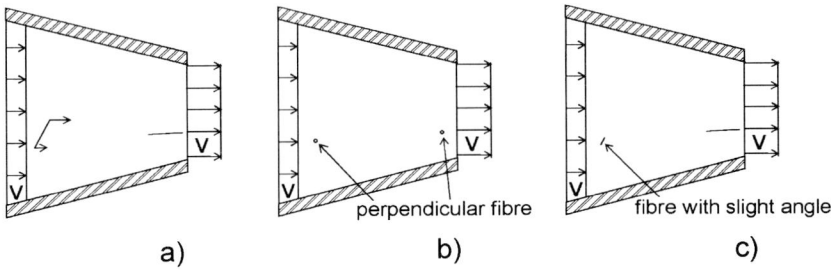

Figure 2.2   *Fibre perpendicular to orientation effect, example with convergent flow. a) fibre orientated by orientational flow b) perfectly perpendicular fibre: no effect c) fibre with slight angle from perpendicular: orientational flow can take effect.*

The orientation effects above are valid only for fibres in the plane where the variation in flow velocity exists, Fig. 2.2a. Fibres perpendicular to this plane (Fig. 2.2b) should not change their orientation, because the velocity over the length of the fibre will be constant. However the resulting orientation in mouldings indicate that the effects are able to orientate perpendicular fibres as well, mainly because perpendicular fibres will not be exactly perpendicular to the orientational effect, Fig. 2.2c. This will be discussed later on in this chapter. Slight deviations from the perpendicular angle, which will usually exist, will enable the orientational flow to take effect.

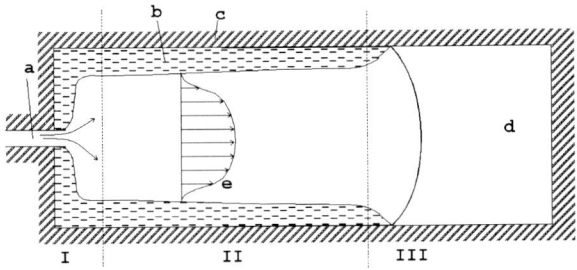

Figure 2.3   *Schematic representation of the filling of the mould. Three regions can be distinguished: I: Mould entrance, II: Transportation zone, III: Flow front. a: gate, b: frozen layer, c: mould surface, d: unfilled part of mould, e: flow velocity profile*

Orientation formation in the mould
The theoretical orientation effects mentioned above will act in the mould, where three regions can be distinguished, see Figure 2.3.

The three regions have their effect on the final fibre orientation, though no direct correlation exists between the orientation layers (skin, shell, core, see Figure 1.1) and the effect of these regions. At the flow front *fountain flow* dominates the orientation, in the transportation zone the *shear* effect is most important, while at the mould entrance a *diverging flow* from the small gate into the wider mould cavity exists.

At the flow front fountain flow transports the polymer from the core to the mould wall. The orientation effect in fountain flow was studied using numerical techniques by Bay and Tucker [Bay,91a and b]. In fountain flow fibres were found to be orientated parallel to flow. However the orientation found by most investigators in the skin layer is a random orientation [Akay,91, Bay,89, Lang,81]. This is caused by the fact that in the core the fibres are generally found to be orientated perpendicular to flow direction, and thus perpendicular to the orientation effect in the fountain flow (As in Fig. 2.2b). As was mentioned before small deviations from the perpendicular state, will finally enable the orientational mechanism to take effect. However, the length over which the orientation mechanism works is only half the thickness of the part. Effect of this will be that the fibres that have a good perpendicular orientation in the core will not be orientated parallel, while fibres with a high deviation from the perpendicular state will be orientated in flow direction. Thus the random deviations from the perpendicular orientation of the fibres in the core, is transferred to a random planar orientation in the skin layer.

Behind the flow front a velocity profile between the frozen layers is present. This velocity profile implies variations in flow velocity, perpendicular to the flow direction. The fibres are orientated therefore in flow direction. This is the main effect that orientates the fibres in the shell layer. Here the same effect is present as in fountain flow, the fibres are originally orientated perpendicular to the orientation effect in the shear flow. Opposite to the orientation in the skin layer, the resulting orientation in the shell layer is quite well aligned to flow direction [Akay,91, Bay,89, Wyzgoski,88, Lang,87b]. The difference with the fountain flow is most likely caused by the fact that the shear flow works for a longer time. In the case of the fountain flow, the fibres can be orientated only while the melt is exposed to the fountain flow. The orientation mechanism has a longer time to take effect to form the orientation in the shell layers (shear flow) than in the skin layers (fountain flow).

It was stated already that the velocity profile will be parabolic if the viscosity is constant. However two effects will change the profile: The temperature over the thickness will not be constant, thus leading to a higher viscosity at locations with lower temperature. The second is the shear thinning effect: the viscosity will be lower, when the shear rate is higher.

The temperature effect leads to a higher viscosity close to the frozen layer. Consequence of this is that the highest shear does not take place directly at the frozen layer. The shear thinning effect leads to a concentration of the shear in a small area of the thickness, Fig. 2.4. The extreme is when all the shear exists close to the frozen layer: plug flow. A second shear thinning effect is due to the energy dissipation in shear flow, this effect must be separated from the real shear thinning effect, which is at constant temperature.

Chapter 2, Orientation

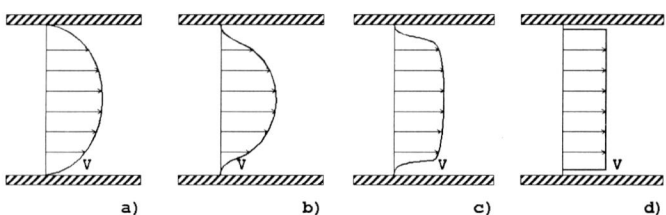

Figure 2.4   *Velocity profiles over the mould thickness. a) Parabolic, b) Including higher viscosity at the frozen layer, c) including high shear thinning effect, d) plug flow.*

The shear thinning effect is dependent on the polymer, Polypropylene for example shows more shear thinning than Polyamide, thus leading to a different velocity profile, more like Figure 2.4c. The result is less orientation in flow direction, in the case of GFPP [Bright,81]. The shear thinning effect tends to concentrate all shear in a very thin layer (If the shear thinning effect is sufficiently strong). Thus it can be the cause of differences in fibre orientations, in the case that different matrix materials are used. The shear thinning effect is also dependent on the fibre volume fraction and length, with a higher shear-thinning at higher fibre fractions and higher fibre aspect ratio [Folkes,80]. The shear thinning effect is important in determining the sensibility of the fibre orientation to the injection moulding conditions. Normally a lower injection speed leads to less shear thinning, and fibre orientation more in flow direction [Bright,81]. The viscosity of the polymer melt is also lowered locally at the location of highest shear, because of the temperature rise due to high energy dissipation. Both shear thinning effect and energy dissipation are most important in one layer, the layer of highest shear deformation. The importance of both effects depends on the matrix polymer type, and on the volume fraction and average length and thickness (aspect ratio) of the reinforcing fibres.

At the mould entrance, the polymer flows from a narrow gate into the wider mould cavity, thus inducing perpendicular orientation of the fibres. The orientation of the fibre before the gate is also of importance, and can be influenced by the runner design.

Once the mould is fully filled, the injection pressure is maintained during cooling of the part. This packing stage partly compensates for the shrinkage of the polymer. Polymer flows into the mould during this stage, leading to an increase in orientation in flow direction.

<u>Orientation variations in the plane</u>
The orientation mechanisms that work to form the orientation layers in the thickness, also work in the plane of the moulding. This leads to a variation in orientation over the mould, as shown in Figure 1.1 for the injection moulded square plate used in this research. Relative thicknesses of the skin, shell and core layers will vary over the plate, as well as the degree of orientation in these layers. Furthermore an effect of shear flow, induced by the sides of the mould, will induce a parallel orientation close to the side of the plate, even in the core layer.
Mostly an increase of fibre orientation in flow direction is found, with increasing distance to the gate [Bright,78]. The same researchers though found little or no change with flow path, at lower injection speed. Hegler found an increase in perpendicular orientation with distance to the gate [Hegler,84]. In the same article Hegler discussed the orientation in a plate with a moulded-in

hole. Downstream from the hole an orientation in flow direction appears, along the weld-line. This principle was applied by changing a line-gate to a multiple-gate (for a square plate) in a study by a student from our University, in cooperation with DSM. The orientation of the fibres in flow direction was increased, in the core layer. This led to an increase in strength (in flow direction) of 13% to 20% [Geerling,93]. Naturally the strength in perpendicular direction was affected negatively.

Deviations from the planar orientation
In most researches the fibre orientation is considered to be planar [Folgar,84, Gadala-Maria,93], all fibres are assumed to have an orientation parallel to the mould or product surface. Some researchers include out of plane fibres in their measurements [Hine,93, Toll,93, Bay,92, Bay,89, Konicek,87] and a considerable degree of out of plane orientation was measured, especially in the core area [Toll,93]. In these cases the measurements were done on polished surfaces, parallel to the mould surface. Fibres with negative and positive angles can not be distinguished [Toll,93], and a random distribution of positive and negative angles is generally assumed.

However, on fractographs of transverse specimens, fibres with an angle to the mould surface were seen, in the core layer [Horst,95a]. These fibres at an angle are only part of the fibres, the rest of the fibres has an orientation parallel to the specimen surface, and perpendicular to the Mould Flow Direction (MFD), which is the generally reported fibre orientation in the core area. This effect of fibres an angle was seen only in thick (5.75mm) samples, and is shown schematically in Figure 2.5.

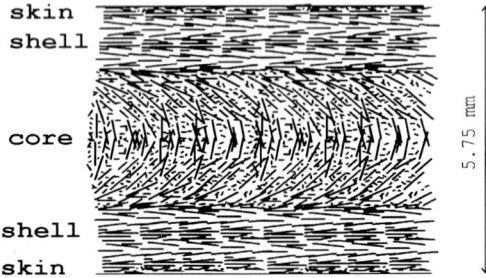

Figure 2.5   *Non planar fibre orientation. Mould flow direction from left to right.*

Close to the shell layer these fibres have a small angle to the surface. This angle increases to perpendicular to the mould surface at the centre of the core. Although it is thought that the influence on the properties of the material are small, it must be taken into account in certain cases. In the tests presented in Chapter 6, where microtomed foils cut from fatigued specimens were testes, the out of plane orientation was found to have considerable influence on the results. Furthermore the existence of these fibres at an angle are not seen to occur in simulations of the mould filling, as the fibre orientation is often assumed to be planar. This indicates that much work has to be done in this area, to improve the predictive value of these simulations.

2.3 Orientation description methods

The orientation is often represented by terms like "parallel to flow", "perpendicular to flow" or "random". For a first impression of the fibre orientation this is correct, of course, though no exact information is given. However for a fibre orientation that seems to be random, often an average direction can be determined. Furthermore for calculations both of orientation in polymer flow, as well as for the calculations of mechanical properties, a more quantitative description of the orientation must be used.

The fibres are aligned in a certain average direction, with a certain scatter around this average direction. Various methods exist to describe this orientation using quantitative parameters. The complete description can be done using the distribution function, in which a three-dimensional fibre-orientation can be described. Because the fibres are often considered to have a planar orientation, the description mostly is 2-dimensional. Especially at complex geometries of moulds it can be necessary to use a 3-D distribution function.
Because the distribution function is quite extensive, certainly for numerical use, often one or more parameters are used to characterize the fibre orientation, often in Tensor notation.

The distribution function
The orientation of an individual fibre can be given by angles θ and ϕ, Figure. 2.6. To the fibre we can assign a unit vector, $\underline{p}$ of which the components can be given by:

$$p_1 = \sin\theta \cos\phi \qquad 2.1a$$

$$p_2 = \sin\theta \sin\phi \qquad 2.1b$$

$$p_3 = \cos\theta \qquad 2.1c$$

Figure 2.6   Coordinates system and definitions of θ, ϕ and $\underline{p}$. [Advani,87]

The distribution function $\psi(\theta,\phi)$ (Advani,87 en Ranganathan,90), is defined thus that the possibility to find a fibre with orientation between θ_1 and $\theta_1 + d\theta$, and ϕ_1 and $\phi_1 + d\phi$ (Or the fraction of fibres with this orientation within a certain volume) equals $\psi(\theta,\phi)\sin\theta_1 d\theta d\phi$. This function ψ must satisfy certain conditions, Firstly that a fibre with angles (θ,ϕ) cannot be distinguished from a fibre with angle $(\pi - \theta, \phi + \pi)$, the distribution function thus is a periodic function:

$$\psi(\theta,\phi) = \psi(\pi - \theta, \phi + \pi) \qquad 2.2$$

$$\psi(\underline{p}) = \psi(-\underline{p}) \qquad 2.3$$

furthermore ψ must be normalised:

$$\int_{\theta=0}^{\pi} \int_{\phi=0}^{2\pi} \psi(\theta,\phi)\sin\theta \, d\theta d\phi = \oint \psi(\underline{p})d\underline{p} = 1 \qquad 2.4$$

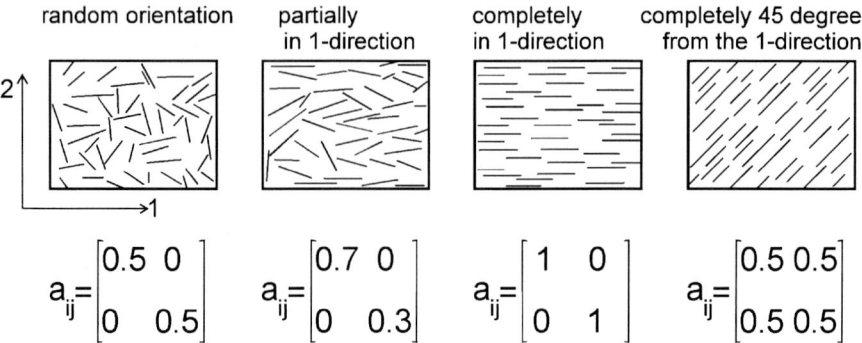

Figure 2.7   Examples of orientations in the plane, with accompanying tensor, two-dimensional. [Ranganathan,90]

Orientationtensors
Advani and Tucker [Advani,87] define orientation tensors, the second and fourth order tensors are respectively:

$$a_{ij} = \oint p_i p_j \psi(\underline{p})d\underline{p} \qquad 2.5$$

$$a_{ijkl} = \oint p_i p_j p_k p_l \psi(\underline{p})d\underline{p} \qquad 2.6$$

Higher order tensors are possible, the odd tensors are all nil, because the distribution function is even. Advani and Tucker [Advani,87] state that the second order tensor describes the fibre orientation nearly as well as the fourth order. Therefore they use the second order tensors in the calculation of fibre orientation in melt flow.

Both the distribution function as well as the orientation tensors can be used to characterise two-dimensional orientation, by using $\theta = \pi/2$. Due to symmetry: ($a_{12} = a_{21}$) and normalisation: ($a_{11} + a_{22} = 1$) two independent elements of the tensor remain, see Figure 2.7. These represent the degree of orientation in the 1-direction (a_{11}), while a_{12} represents the deviation of the orientation-direction from the 1-direction.

Orientation parameters
Pipes et al. [Pipes,82] takes the 2-dimensional case, and calls his distribution function n(φ). The same symmetry and normalisation condition as in the 3-D situation apply. He defines his orientation parameters as shown below:

$$f_p = 2<\cos^2\phi> -1 \qquad 2.7$$

$$g_p = [8<\cos^4\phi> -3]/5 \qquad 2.8$$

$$<\cos^m\phi> = \int_0^{\pi/2} n_p(\phi)\cos^m\phi \, d\phi \qquad 2.9$$

The subscript p denotes that this is for the planar orientation case. For the axial case (fibres are orientated in cylinders) the constants change. In many articles [Darlington,91, Matsuoka,90, Voss,88] only one parameter is used, f_p from equation 2.7. This parameter varies between 1 and -1 in the case of perfect orientation in the 1-direction, respectively perpendicular to it. For random in the plane orientation f_p equals 0.

2.4 Orientation measurement

Various methods have been presented to measure the fibre orientation in a product or specimen. Roughly these methods can be divided into two categories. The first is a surface method, where the orientation is studied from a polished surface. The second is a method using thin films cut from the specimen. The fibre orientation in these films is made visible by transmitting light or X-rays through the film.

In the method using polished sections of the specimen, the angle and ellipticity of the resulting ellipses for each fibre are measured. Automated procedures can be used to measure a high number of fibres in a short time. The method can be used to measure both in-plane orientation, as well as 3-D fibre orientation [Hine,93, Toll,93, Bay,92, Bay,89, Konicek,87]. Hine et al. concluded that fibre angles smaller than 10° could not be accurately measured. As this will give problems if the cross section is taken perpendicular to the fibre orientation, this problem is overcome by taking the cross section at an angle to the main fibre direction. Another problem is the phenomenon that fibres perpendicular to the polished section have a higher probability to occur at the surface, than fibres at an angle or fibres parallel to the surface. A weighting factor has to be introduced, to compensate for this [Bay,89].

A more general view of the orientation of the fibres can be had by making a thin slice of the specimen, and using light (Transmission Optical Microscopy) or X-Ray (contact microradiography [Delpy,85, Folkes,82, Folkes,80, Bright,78]) to make a print of the orientation [Folkes,82]. This will be a general view, indicating the thicknesses of the layers and roughly the orientation in these layers. Only the in-plane orientation can be obtained. Gadala-Maria (1993) used a method to automatically measure the orientation of single fibres, in thin films. A problem is the contrast between fibres and matrix. Special tracer fibres can be used for this.

A similar rough impression of the fibre orientation and thickness of the respective layers can be obtained by studying fractographs [Horst,95a], or low magnifications of polished sections [Akay,91, Ranganathan,90, Delpy,85, Kaliske,73].

2.5 Orientation prediction

As was indicated in the introduction, the perfect situation will be if the fibre orientation in products can be accurately predicted *before the mould is manufactured*. Various publications are available in this field [Baraldi,92, Frahan,92, Bay,91a, Bay91b, Altan,90, Matsuoka,90, Advani,87, Vincent,86]. Two basic methods have been used. For both methods the *polymer flow field* is *calculated first*. In the first method general rules give the degree of orientation caused by a certain flow field. The second method follows individual fibres during their path through the mould. In both cases especially fibre-fibre interactions are difficult to model, and complicate calculations, especially at higher volume fractions of fibre. The fibres also change the rheological behaviour of the polymer, giving a higher viscosity at low shear rate, while at higher shear rates the viscosity remains the same. Thus the shear thinning effect is increased [Folkes,80]. This effect can of course be included in the viscosity that is used for the flow simulation. However the *local viscosity* will be dependent on the *local fibre orientation* as well. By doing the flow simulation first, and consequently the fibre orientation analysis, this effect can never be accounted for.

Fibre orientation directly from flow field calculations
Various commercial mould fill simulation software programs are available, with a capability to predict the fibre orientation. Results of these predictions are consistently poor [Baraldi,92]. None of these rely upon specific calculation modules performing a real orientation analysis. Normally after completion of the flow simulation, the computed velocity and temperature profiles lead to qualitative conclusions of the fibre orientation. Mostly the 3D FEM flow simulation softwares provide information of velocity fields in two modes:
In mode 1 the average velocity vectors over the thickness can be viewed, at different locations, and at different times during the filling. Thus the presence of converging or diverging flow can be located, and at locations with homogeneous flow the flow direction (mostly coinciding with fibre orientation direction in the shell layers) can be seen.
In mode 2 the velocity profiles over the mould thickness at certain locations and certain stages during filling can be viewed. This can give some idea of the relative thicknesses of the core, shell and skin layers [Vincent,86].

Originally this method to determine the fibre orientation from the flow simulation was only valid for dilute suspensions of fibres (low fibre fractions). However the Fibre Interaction Parameter first introduced by Folgar and Tucker [Folgar,84] was devised to compensate for this. It is based on the assumption that fibre-fibre interactions *randomize the fibre orientation*, preventing the fibres from assuming a perfect alignment in steady-state flow. The fibre interaction parameter can not be calculated, and has to be measured experimentally. The method gives good agreement with experimental results [Folgar,84], and is used successfully by various authors [Bay,91a and b, Matsuoka,90].
Still, flow simulation softwares offer *mainly qualitative predictions* of the fibre orientation. The exact level of orientation and thicknesses of layers can not be calculated. Orientation at complex situations like dividing or merging flows and weld lines are not available. Mostly the contribution to the orientation that was made by the fountain flow at the flow front is fully ignored. Bay and Tucker [Bay,91a] made a comparison of fibre orientation predictions with and without the inclusion of the fountain flow effect. Without fountain flow the orientation near the surface was as aligned as in the shell layer. Including the fountain flow led to a more random orientation near the surface, which is normally seen in products as well.

Following fibres along the flow path
This method does not differ from the former in one aspect: First the flow path is calculated, using behaviour of homogeneous polymers. Then a set of initial positions and orientations is selected at the flow entrance. Fibre trajectories are determined from the known velocity fields. As the distance between the trajectories initiating at the different fibre ends can change from the initial distance (the fibre length), the fibre has to be repositioned to the middle of the connecting line. These steps have to be repeated until the fibre reaches its final position [Cengiz Altan,90, Baraldi,92]. This procedure provides more convincing results for the evaluation of orientation at complex flows.

2.6 Conclusions

The formation of orientation-formation is a complex phenomenon, in which many parameters are of influence. For simple geometries like square plates containing a line gate the fibre orientation varies through the thickness, and from place to place throughout the surface. This causes anisotropy and inhomogeneity of physical and mechanical properties. The complexity of the fibre orientation is enormous, and can never be described in one or a small number of parameters. However use of the distribution function, which describes the orientation in full, is practically impossible. The distribution function is a function of not only θ and ϕ, but also of the coordinates x, y, z, in the specimen or product.

Prediction of the fibre orientation using FEM calculations is possible, but still suffers from lack of accuracy. Especially for complex shapes, where the fibre orientation varies strongly over a small distance, predictions will not be accurate. However the expectation is that quality of predictions will increase, due to both increase in knowledge of the fibre orientation mechanisms and especially fibre-fibre interactions, and the increase in calculation speed on modern computers.

In Figure 2.8 photographs of a sectioned product containing ribs is shown, the complexity of fibre orientation visible here is an example why the fibre orientation predictions will need to be very accurate. Moreover because these corners provide stress concentrations, and therefore possible locations for failure initiation.

One last shortcoming of the available algorithms for the prediction of the fibre orientation is that the orientation is often presumed to be an *in the plane* orientation. As was shown in § 2.2 this need not be the case. Especially for products containing relatively thick sections, an error is made if planar orientation is assumed.

16   Influence of fibre orientation on fatigue of short glassfibre reinforced Polyamide

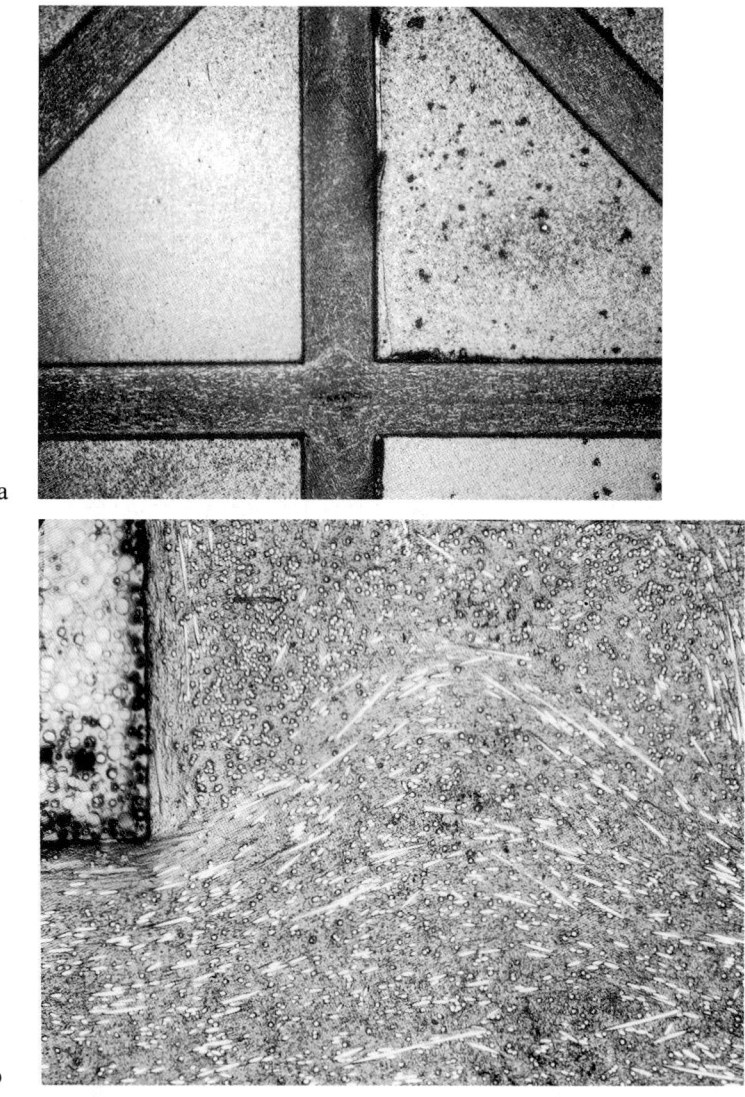

Figure 2.8   *Orientation at location of rib, flow direction from left to right. b is detail of a*

3. Mechanical behaviour of composites

3.1 Introduction

The fibre orientation that is the result of the mould filling process, gives rise to variations of physical and mechanical properties in the final product. Physical properties that are of importance are the thermal conductivity and the thermal expansion. The thermal conductivity increases with increasing degree of fibre orientation (in the direction of orientation), while the thermal expansion decreases. Both are important features, especially in cooling of injection moulded parts, and are caused by different thermal conductivity and expansion of fibres and polymer matrix. The thermal expansion coefficient correlates with the fibre orientation [Matsuoka,90], and can be used as a measure for the orientation state.

The mechanical behaviour is also greatly influenced by the fibre orientation, and is the objective of the research presented here. To what extent does the fibre orientation affect the mechanical properties, and in particular the fatigue behaviour? As was shown in the previous chapter the fibre orientation changes throughout parts, and therefore the mechanical behaviour. In case the stiffness in tension is concerned, the resulting behaviour of the *part* will be an average over the length of the part of the local properties. The Elastic Modulus therefore is a *global property*. For the tensile strength of a part the minimum value in the length of the specimen will apply, which is easily seen in the tests performed. As the strength of specimens cut from the square plate under study varies over the length of the specimen, the specimen will always break at about the same location, at the weakest spot. The strength therefore is a *local property*. This can be seen in the fibre orientation as shown in Figure 1.1. Further from the middle of the plate the angle of the fibres in the Core layer with respect to the axis of a transverse specimen increases. Thus the strength decreases, making transverse specimens fracture at one end, never in the middle. For the case where specimens are subjected to bending, the situation changes of course, and the stress distribution over the thickness due to the bending moment must be considered as well.

Through thickness also an averaging procedure for the stiffness properties applies. The orientation varies over the thickness, resulting in layers of higher and of lower stiffness. In axial loading the strain over all layers must be the same, thus the stresses in different layers will be different. Integration of the stresses in the layers over the thickness will give the total loading, at a given strain. The average stress in the specimen will be found by dividing by the thickness. For the strength also a "weakest link" applies, with the difference that if one layer fails, other layers will take up the load and will either fail, or prevent failure. Hitchen describes that cracks in the core layer (fibres perpendicular to the loading direction) exist, while the specimens continue to resist fatigue loading [Hitchen,93a and b]. Therefore I would like to distinguish the weakest layer from the critical layer. The *weakest layer* is the layer that fails first, not necessarily accompanied by failure of the entire specimen. The *critical layer* is the layer that dominates the failure of the specimen. Weakest layer and critical layer may be the same.

In this chapter a short overview will be given of the mechanical properties of fibre reinforced plastics, in relation to the fibre orientation. The theory of fibre reinforcement will be discussed. Using the theories for calculating the stress in single short fibres, the stiffness of a short fibre composite will be discussed. Theories for calculating strength will be treated as well. Calculation of impact strength, through considerations of the energy dissipation mechanisms in fracture are also possible, but will not be discussed here, as impact properties were not considered in the current research. An introduction can be found in [Horst,93] or [Piggott,80]. Experimental results of Elastic Modulus and strength from literature will be compared to predictions using the fibre orientation.

Finally experimental results for the specimens under study will be given. The effect of strain rate and humidity on the specimen tensile strength is discussed.

3.2 Theoretical evaluation of Modulus with fibre orientation

A lot of literature is available on the theoretical determination of mechanical properties of short fibre reinforced composites [Piggott,80, Folkes,82, Agarwal,90]. In this paragraph methods to calculate the most relevant properties (stiffness and strength) will be discussed. For each of these first behaviour for the case of aligned fibres parallel to the loading direction will be treated, then aligned fibres perpendicular to loading and finally fibres at an angle to the load.

3.2.1 Theoretical determination of Elastic Modulus in fibre direction

For the determination of the stiffness of a short fibre unidirectional composite, first the stress distribution in a single fibre is discussed, using elastic theory. Using the average stress in the fibres, an effective Modulus can be calculated. Consequently the fibre stress in the case of non-elastic matrices will be given, and finally the Elastic Modulus in directions other than along the fibre will be treated.

Stress distribution in a fibre (elastic theory)
When a composite with perfectly aligned fibres is loaded in the fibre direction, the load is transferred from the polymer matrix to the relatively stiff fibres, through shear stresses on the interface. The calculation of the distribution of shear and tensile stresses along the fibre was first done by Cox [Cox,52], for fully elastic matrix and fibre, and perfect bonding.

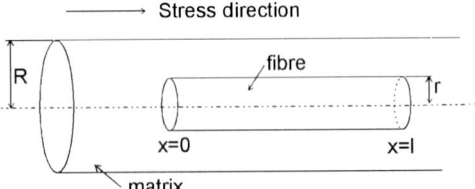

Figure 3.1   A single fibre in the matrix [Folkes,82]

A fibre of length l and radius r is surrounded by matrix. The strain in the matrix is assumed to be homogeneous. The shear stress along the fibre and the tensile stress inside the fibre will vary over the fibre length. Cox uses a force balance over fibre length dx, as in equation 3.1

Chapter 3, Mechanical behaviour of composites

$$dF = \tau(x)\, 2\pi r\, dx \quad \text{or:} \quad \frac{dF}{dx} = 2\pi r\, \tau(x) \qquad 3.1$$

According to the assumption both matrix and fibre are elastic, so shear stress τ is proportional to shear strain γ, which in turn is proportional to the difference between displacement u in the fibre and v in the matrix. Using this the distribution of stress over the fibre can be calculated:

$$\sigma_f = \frac{F}{A_f} = E_f\, \epsilon \left\{ 1 - \frac{\cosh \beta(\frac{l}{2} - x)}{\cosh \beta \frac{l}{2}} \right\} \qquad 3.2$$

The fibre is assumed linear elastic, with Modulus E_f. ϵ is the strain in the matrix. The average stress in the fibre, of main interest for the calculation of the stiffness, is: (eq. 3.3)

$$\bar{\sigma}_f = E_f\, \epsilon \left\{ 1 - \frac{\tanh \beta(\frac{l}{2})}{\beta \frac{l}{2}} \right\} \qquad 3.3$$

$$\beta = \sqrt{\frac{H}{E_f A_f}} \qquad 3.4$$

H is a constant, depending on Modulus of elasticity of fibre and matrix, and geometry.

3.2.2 Calculation of the effective Modulus

The average stress in fibre direction, for a composite with fibre fraction v (volume fraction) can be calculated as the average of stresses in fibre and matrix: (3.5)

$$\sigma_0 = v_f\, \bar{\sigma}_f + (1 - v_f)\sigma_m \qquad 3.5$$

σ_m, the stress in the matrix can be written as $\sigma_m = E_m\, \epsilon$, using the Modulus of the matrix E_m. This gives:

$$\sigma_0 = v_f\, E_f\, \epsilon \left\{ 1 - \frac{\tanh \beta \frac{l}{2}}{\beta \frac{l}{2}} \right\} + (1 - v_f) E_m\, \epsilon \qquad 3.6$$

So the longitudinal Modulus of the short fibre composite is given by 3.7:

$$E_0 = \frac{\sigma_0}{\epsilon} = v_f\, E_f \left\{ 1 - \frac{\tanh \beta l/2}{\beta l/2} \right\} + (1 - v_f) E_m \qquad 3.7$$

If the distance between fibres equals R_v-r, Cox showed that constant H equals:

$$H = 2\pi \frac{G_m}{\ln\left(\frac{R_v}{r}\right)} \quad \text{and thus:} \quad \beta = \left\{\frac{2\pi G_m}{E_f A_f \ln\left(\frac{R_v}{r}\right)}\right\}^{\frac{1}{2}} \quad \quad 3.8$$

in which G_m is the Shear Modulus of the matrix. R_v is the only remaining unknown factor, and can be solved using the fibre fraction, and a square or hexagonal distribution of fibres. For a hexagonal distribution, R_v can be solved from: $v_f = \dfrac{2\pi r^2}{\sqrt{3} R_v^2}$

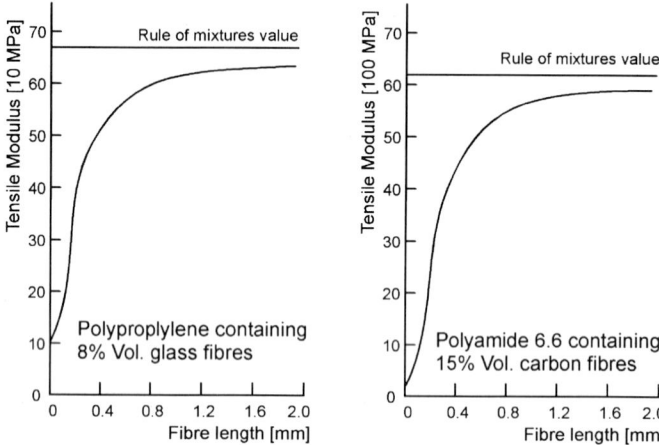

Figure 3.2   Theoretical dependence of longitudinal Modulus on fibre length, linear elastic theory. The rule of mixtures value is the value for continuous fibres. [Folkes,82]

So for any composite, the constant β can be determined. β determines the degree of dependence of the longitudinal Modulus E_0 on the fibre length l. In Figure 3.2 the calculated Modulus is shown for 2 different composites as a function of fibre length. The straight line is the value for continuous fibres.

3.2.3. Stress distribution for the non-elastic case

Cox [Cox,52] assumes elastic behaviour of both fibre and matrix. However the shear stress will have a finite maximum value. The shear stress at the fibre ends will not raise above the shear yield stress τ_y. Different models have been used for τ, of which the simplest is rigid - plastic, introduced by Kelly and Tyson in 1965. [Kelly,65, Tyson,65] For this case the stress in the fibre and the shear stress are depicted in Figure 3.3. The critical fibre length is in this theory the minimum length of the fibre to attain it's maximum (fracture) stress: $l_c = \dfrac{\sigma_{uf} r}{\tau_y}$.

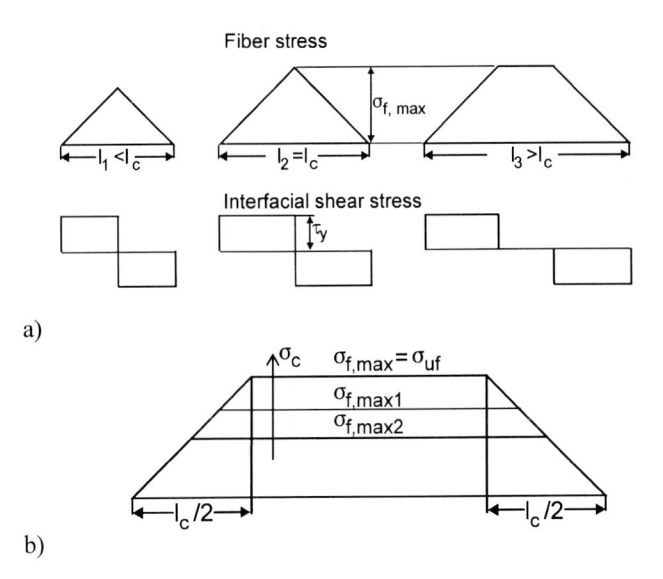

Figure 3.3 a: Variation of the tensile stress in the fibre and the shear stress at the fibre-matrix interface with fibre length. [Agarwal,90]
b: Influence of the composite stress on the stress in fibres longer than the critical length.

Various more elaborate models exist for the calculation of the stress distribution in the fibres, using different material behaviour and possibly using Finite Element Methods (FEM). Whatever the model for the stress distribution in the fibres, the dependence of E_0 on fibre length l will be of the form depicted in Figure 3.2, according to the elastic analysis. All differences can be seen as a change in the effective value of β.

3.2.4. Elastic Modulus in not longitudinal directions

The transverse Modulus E_{90} and the shear Modulus G for unidirectional composites is not easily derived. These properties are matrix-dominated and can be estimated by: [Folkes,82]

$$\frac{1}{E_{90}} = \frac{v_f}{E_f} + \frac{(1-v_f)}{E_m} \quad ; \quad \frac{1}{G} = \frac{v_f}{G_f} + \frac{(1-v_f)}{G_m} \qquad 3.9$$

In glassfibre or carbonfibre reinforced polymers $E_f \gg E_m$ and $G_f \gg G_m$ and thus

$$E_{90} \approx \frac{1}{(1-v_f)} E_m \quad \text{and} \quad G \approx \frac{1}{(1-v_f)} G_m \ .$$

Elastic Modulus for not-unidirectional composites
Various methods have been reported to determine the Elastic Modulus for composites in which the fibres are not fully aligned. [O'Donnell,94, Folkes,82, Lees,68a] The relative efficiency of fibres in each orientation direction can be evaluated, and consequently averaged for the whole

composite. In a continuous fibre composite with aligned fibres, at an angle θ to the applied load, the orientation efficiency factor η_o is introduced. $\eta_o = 1$ for fibres in load direction ($\theta=0$), and $\eta_o = 0$ for perpendicular fibres ($\theta=90°$). For fibres at an angle θ: $\eta_o = \cos^4\theta$. It is assumed that strain in both fibre and matrix are equal (Voigt average).

$$E = \eta_o \eta_l v_f E_f + (1-v_f)E_m \qquad 3.10$$

For non-unidirectional composites η_o is determined by dividing the fibres in groups having the same orientation. The orientation efficiency factor is than summed over all groups:

$$\eta_o = \sum_n a_n \cos^4\theta_n \qquad \text{and:} \quad \sum_n a_n = 1 \qquad 3.11$$

a_n is the fraction of fibres at an angle θ_n to the loading direction. Of course the summation can be replaced by an integration, in case the fibres can not be grouped.

In eq. 3.10 the overall efficiency factor $\eta = \eta_o\eta_l$ is used with η_o the orientation efficiency, and η_l the efficiency according to the length of the fibre, which equals $\eta_l = 1 - \dfrac{\tanh \beta l/2}{\beta l/2}$ if Cox's elastic theory is used. If not all the fibres are of equal length either an average has to be used, or a summation or integration over the different fibre lengths.

An other method to calculate the stiffness if the composite is not unidirectional was introduced by Halpin and Tsai. This method uses the theory commonly used for laminates, in which the stiffness of a plate is assumed to be the average of the stiffnesses of all layers:

$$E = \sum_n \frac{t_n}{t} E_n; \quad t = \sum_n t_n;$$ in which E_n is the stiffness of layer n having thickness t_n. In most short-fibre reinforced composites the orientation can be seen as a layered structure, see Chapter 2, so this method is widely employed. Difference with the method explained before, is that no specific method for determining the stiffnesses E_n is assumed.

3.3. Theoretical prediction of strength

For the determination of the strength of a composite, first strength in case of aligned fibres in loading direction will be determined. Two possibilities are distinguished: fibres exceeding the critical length l_c, and fibres shorter than the critical fibre length. Consequently the strength of composites with fibres not in loading direction will be discussed.

3.3.1 Strength of unidirectional, short-fibre composites

Prediction of the strength of short fibre composites is a complex problem, even in the ideal situation of aligned fibres in the loading direction. Failure can occur either in the fibres, the matrix, or at the fibre-matrix interface. The generally used carbon and glass fibres have a small strain at failure, compared to that of the matrix polymer. If the fibres are well-bonded to the

Chapter 3, Mechanical behaviour of composites

matrix, and their length exceeds the critical length l_c, the failure stress of the composite can be given by equation 3.12,

$$\sigma_{uc} = v_f \sigma_{uf} + (1-v_f)\sigma_m' \qquad 3.12$$

in which σ_{uf} is the failure strength of the fibres, and σ_m' is the stress in the matrix, at the failure strain of the fibres. In practice for short fibre composites this value for continuous fibre reinforced composites is not attained. The average stress in the fibres at failure is lower than the value σ_{uf}. In the Kelly and Tyson model for the fibre stress, see section 3.2.3, the average fibre stress can be calculated. The strength of the composite is:

$$\sigma_{uc} = v_f \sigma_{uf} \left(1 - \frac{l_c}{2l}\right) + (1-v_f)\sigma_m' \quad \text{for } l>l_c \quad \text{and} \quad l_c = \frac{\sigma_{uf} r}{\tau_u} \qquad 3.13$$

τ_u is the shear strength of interface or matrix. l_c is the critical fibre length.
For fibres shorter than the critical length, the interface fails, not the fibre. The maximum fibre stress is given by: $\tau_u l/r$, the average fibre stress is half this value. The strength of the composite is:

$$\sigma_{uc} = v_f \frac{\tau_u l}{2r} + (1-v_f)\sigma_m' \qquad \text{for } l<l_c \qquad 3.14$$

In actual composites a distribution of fibre lengths will exist, making summation or integration over the fibre lengths necessary.

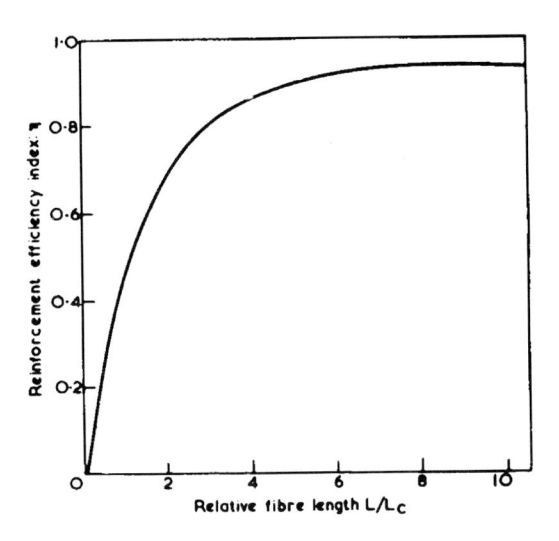

Figure 3.4   Dependence of the reinforcement efficiency index η_l on fibre length. $\eta_l = 1$ for continuous fibres. [Folkes,82]

The theoretical strength for a unidirectional composite with fibres all of equal length is shown in Figure 3.4.

It is clear that the critical fibre length l_c has a considerable effect on composite strength. l_c can be determined if σ_{uf}, r and τ_u are known. For glassfibre reinforced Polyamide 6.6 $l_c \approx 230$ μm. Both r and τ_u can be used to improve the strength. By using glassfibres with a small diameter, and good adhesion to the matrix, the fibre breakage during compounding and injection moulding can be compensated for to some extent. However, we must realise that τ_u is the minimum value of both the interface and the matrix shear strength. In the case of GFPA 6 used in this research, the *interface shear strength is higher than the matrix shear strength*. The matrix fails, not the interface in case of a tensile test. Improvement of interface strength in this case has no influence on composite strength.

3.3.2. Strength in non-fibre direction

As is the case for the stiffness, also the strength of a composite decreases if the angle between fibres and loading direction increases. The strength of a composite perpendicular to the fibre direction can be lower than the pure matrix strength. This is caused by the stress-concentration effect of the fibres. The anisotropy of strength can not be easily determined theoretically, because the composite can fail in various modes. Stowell and Liu [1961] consider three:

I) For fibres in the loading direction and for small θ, failure is determined by fibre strength, see 3.3.1.
II) With increasing θ the shear stresses in both matrix and interface increase. The failure will be determined by shear.
III) When θ approaches 90° the mechanism changes into a tensile mechanism, perpendicular to the fibres. Failure can occur either in the matrix or the interface. Strength can be lower than the strength of the pure matrix, due to the stress concentration effect of the fibres.

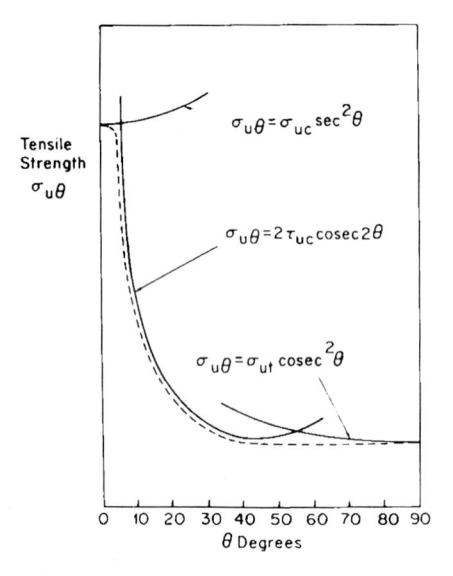

Figure 3.5   *Calculated strength-anisotropy for a unidirectional composite, based on Stowell and Liu's theory. [Folkes,82]*

Chapter 3, Mechanical behaviour of composites

The Stowell and Liu equations for the failure stress at an angle between fibres and loading direction θ, for the three failure modes is given by: (in brackets the notation used in Fig. 3.5)

I) $\sigma_{u\theta} = \sigma_{uc} / \cos^2\theta \ (= \sigma_{uc} \sec^2\theta)$

II) $\sigma_{u\theta} = \tau_{uc} / \sin\theta\cos\theta \ (= 2\tau_{uc} \operatorname{cosec}2\theta)$ 3.14

III) $\sigma_{u\theta} = \sigma_{ut} / \sin^2\theta \ (= \sigma_{ut} \operatorname{cosec}^2\theta)$

σ_{uc} and σ_{ut} represent the strength of a unidirectional composite in fibre direction, respectively perpendicular to this. τ_{uc} is the in-plane shear strength. $\sigma_{u\theta}$ is shown schematically in Figure 3.5.

Other failure criteria exist, among which energy-based criteria. These will not be presented here, as only an introduction to the theoretical determination of mechanical properties was intended.

In injection moulded components the fibres will not be unidirectional, furthermore the orientation will vary through the component. For an estimate of the strength a similar approach can be used as was discussed for the stiffness. For the special case of random-in-the-plane fibres the Stowell Liu equations (3.23) can be integrated over the angle θ, between 0° and 90°, giving a value for the average strength of the composite:

$$\bar{\sigma}_{uc} = \frac{2\tau_{uc}}{\pi} \left\{ 2 + \ln\left(\frac{\sigma_{uc}\sigma_{ut}}{\tau_{uc}^2}\right) \right\}$$ 3.15

3.4 Comparison of properties with predictions

Some literature is available in which actual predictions of properties, as dependent on fibre orientation, is compared to values measured on real systems. To show the possibilities in this field, some comparisons will be discussed here. Predictions are largely based on the theories summarised in the former paragraphs. The stiffness will be treated first, followed by results for the strength.

3.4.1 Stiffness

For aligned fibres, tested at an angle to fibre direction, quite some work is known from continuous fibre studies. For short fibre composites manufacturing of specimens containing aligned fibres is one of the main reasons why results cannot be readily compared to injection moulded specimens.

Angular dependence of tensile Modulus for aligned composites was discussed by McNally [McNally,77], who compares experimental results from various authors. The formula for angular dependence of the tensile Modulus is not using the Moduli of fibres and matrix, as in Equation 3.10. The moduli in direction of alignment, respectively perpendicular to this are used. Equation 3.16 is the Lees model, while the more complex Shaffer model is shown in equation 3.17.

$$E_\theta = \frac{E_L \cdot E_T}{E_L + \cos^4\theta \, (E_T - E_L)} \qquad 3.16$$

E_L and E_T are the Modulus Longitudinal respectively Transverse to "fibre direction", in this case the Mould Flow Direction MFD. The angular dependence is shown in Figure 3.6 for polyethylene/33% glassfibres and styrene-acrylonitrile/20% glassfibres.

$$\frac{1}{E_\theta} = \frac{\cos^4\theta}{E_L} + \frac{\sin^4\theta}{E_T} + \left\{ \frac{1}{\tau_{XY}} - \frac{2\nu_{XY}}{E_T} \right\} \sin^2\theta \cdot \cos^2\theta \qquad 3.17$$

In McNally's experiments on Glassfibre reinforced PBT (30%Wt. Glassfibres), specimens with aligned fibres are made using compression moulding of extruded strands. The quality of these specimens was generally poor (high void content) and a large number was discarded. Measurements of elastic Modulus (Figure 3.7) were compared with both the Lees (Equation 3.16) and Shaffer model (Equation 3.17).

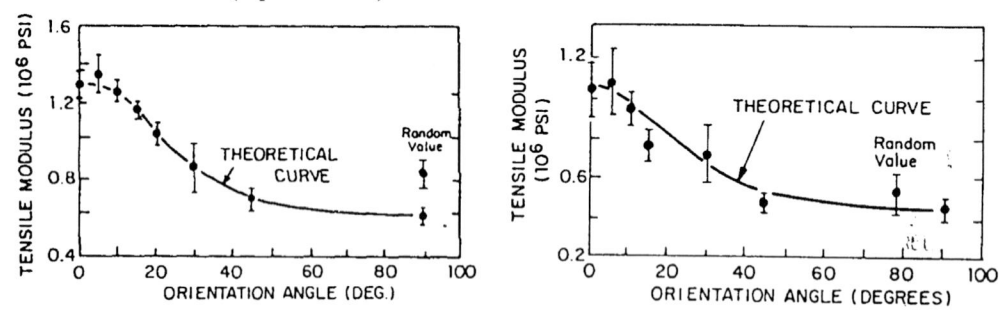

Figure 3.6   Angular dependence of tensile Modulus for aligned composites of left: polyethylene/33% glassfibres and right: styrene-acrylonitrile/20% glassfibres [Lees,68a].

The simpler Lees model describes the angular variation of Modulus as well as the more complex Shaffer one, and should be convenient for use in part design. The Modulus value for random orientation shown in Figure 3.7, can be calculated using the method of Lees:

$$E_{random} = \frac{2}{\pi} \int_0^{\pi/2} E_\theta \, d\theta \qquad 3.18$$

An entirely different approach was presented by Brockmüller and Friedrich in 1992. A micromechanical model was presented and calculations on this model were done using Finite Element Models. The model assumes aligned fibres, and in the FEM calculation an eighth part of the fibre and matrix was modelled, to calculate the stress as a function of strain. Without going into the details of the calculations [Brockmüller,92], the results are shown in Figure 3.8. The experiments were carried out on injection moulded bars. These were assumed to have high fibre alignment. The Elastic Modulus can be predicted quite accurately, although it must be said that a fibre-configuration parameter had to be included in the model. Variation of this parameter results in a variation of about 10% in the calculated values, making it possible to adjust the

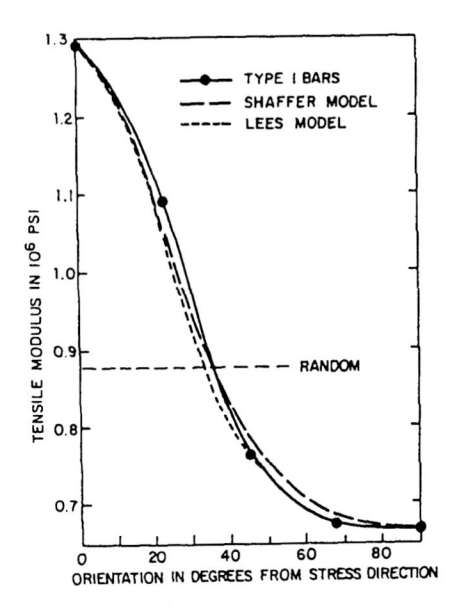

Chapter 3, Mechanical behaviour of composites 27

Figure 3.7 Angular dependence of tensile Modulus for an aligned composite PBT containing 30%Wt. glassfibres. Experimental results (continuous line) compared with Lees and Shaffer models. [McNally,77]

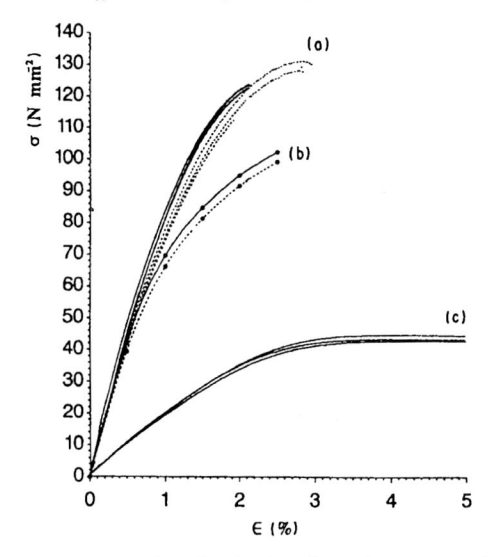

Figure 3.8 Experimental and calculated strain curves. a) Experimental curves for GFPA containing 18 and 14 %(Vol) of glassfibres (Continuous respectively broken lines). b) Calculated values for these materials, c) Experimental curves for pure PA6 [Brockmüller,92].

calculated value of Modulus to the experimental one. Discrepancies between calculations and experiments are considerable though when strength is concerned. Brockmüller states that the Von Mises criterion was used in the model, which is not experimentally confirmed for polymers, and may be the explanation for the differences.

Only one article was found where calculations were based on actually measured fibre orientations in an injection moulded part, and compared to measurements of stiffness. This work was done by O'Donnell and White in 1994. Measurements of the variation of Young's Modulus with depth have been made using a three-point bend test on glassfibre reinforced nylon 6.6 injection moulded bars. A layer removal-technique was used in which on both sides simultaneously a layer was removed. Thus a symmetrical distribution of stiffness over the thickness was assumed. The Halpin-Tsai equations (similar to eq. 3.10) were used to estimate the depth varying Modulus, using the fibre orientation distribution measured at various depths. See [O'Donnell,94] for a detailed account. In Figure 3.9 a comparison is made between calculations and measurements.

The agreement is quite good, except near the surface, also the effect of fibre length seems to be secondary to the effect of fibre orientation. The fall in stiffness below the surface is explained by changes in *matrix stiffness* in this region, not by fibre orientation! The calculations shown in Figure 3.9 used a varying matrix stiffness, leading to results which compared better to the measurements than in the case where a constant matrix stiffness was assumed.

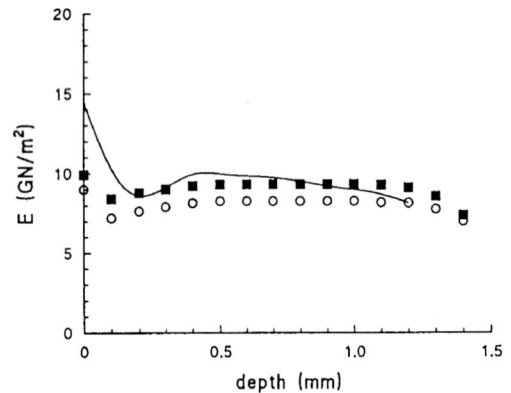

Figure 3.9   Comparison of Young's Modulus distributions. The line is the result of layer-removal bending tests, the points are from the calculations. The upper bound uses a fibre aspect ratio of 60, the lower uses 10 [O'donnell,94].

3.4.2 Strength

For the strength McNally also mentions some examples. The equations used for the strength are those in Equation 3.14. In Figure 3.10 the calculated values using eq. 3.14 and measured values on polyethylene/33% glassfibres and styrene-acrylonitrile/20% glassfibres are shown. In the graph $\sigma_{//}$ and $\sigma_{\perp}$ are used as values for the measured longitudinal respectively transverse strength, and τ for the shear strength of the composite. The three cases are related to the changing mechanism, from tensile failure in fibre direction, to shear at intermediate angles, and tensile failure transverse to the fibre direction.

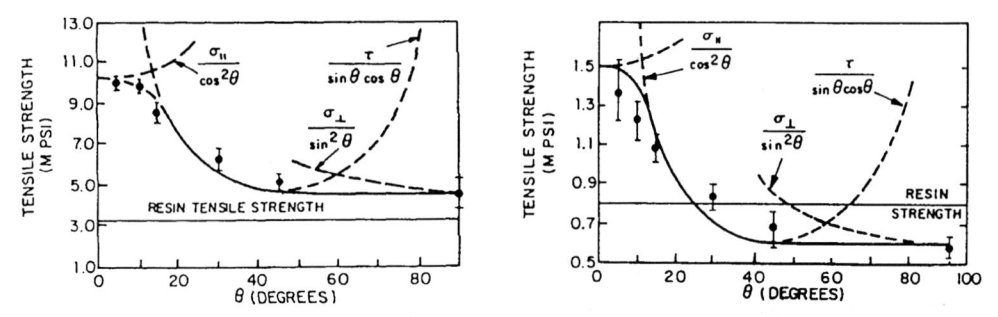

Chapter 3, Mechanical behaviour of composites 29

Figure 3.10   Angular dependence of tensile strength for aligned composites of left: polyethylene/33% glassfibres and right: styrene-acrylonitrile/20% glassfibres [Lees,68b].

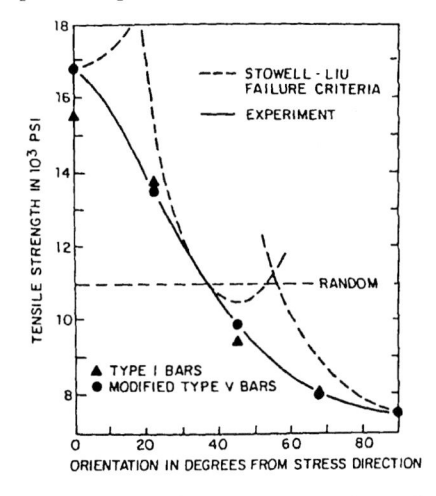

Figure 3.11   Tensile strength variations with angle, for an aligned composite of PBT containing 30%(Wt.) of glassfibres [McNally,77].

The results for McNally's specimens of PBT containing 30%(Wt.) of aligned glassfibres, are shown in Figure 3.11. The Stowell-Liu equations compare quite well to the measured values. Graphical integration yielded the strength value for a randomly orientated composite.

3.5 Mechanical Properties of Specimens used in this research

For the investigations of orientation-influence on tensile strength, one type of injection moulding was used, with some slight variations. In all cases square plates were used, injection moulded through film gates. The dimensions of the plates were 100x100 mm with thicknesses of 2 and 5.75 mm, and 90x90 mm with thicknesses of 2 and 6 mm. Injection moulding conditions for these are summarised in Appendix I. From these plates specimens were milled as shown in Figure 3.12, to obtain values for the mechanical properties. The specimens were tested in tension in a servo-hydraulic MTS type 810, at a constant cross-head speed of 50mm/min. The tensile strength will be discussed here, as this is used extensively in the investigations for comparison with the fatigue strength.

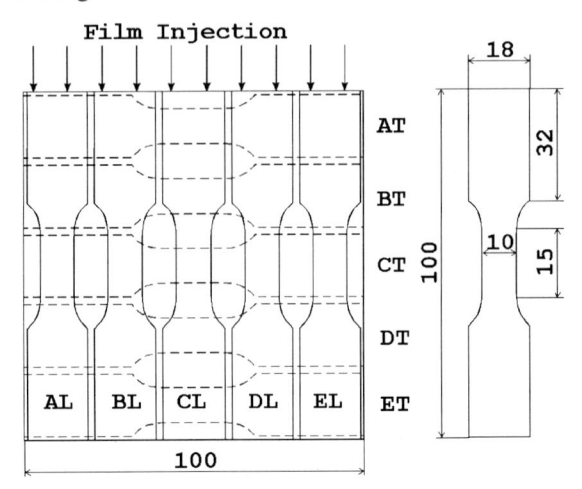

Figure 3.12 Location of specimens in the plate, dimensions of milled specimen. MFD from top to bottom.

3.5.1 Influence of orientation

For conditioned (to equilibrium water content in air) specimens the values of tensile strength are shown in figure 3.13 and 3.14 for 2 and 5.75 mm thick specimens respectively. For the specimens in longitudinal direction the specimens cut from the side of the plate show higher strength, compared to the middle ones. This is caused by differences in the fibre orientation of the core layer: in the middle of the plate (CL) the fibre orientation in the core is perpendicular to MFD. Nearer to the sides of the mould, not only does the thickness of the core layer decrease, but also the orientation becomes more aligned with MFD. Near the side the core layer has

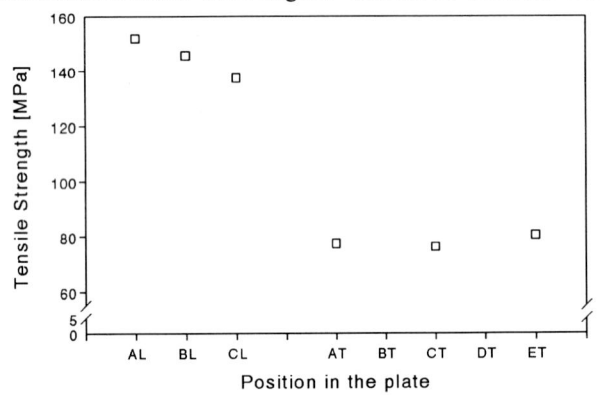

Figure 3.13 Variations of tensile strength over the plate, 2 mm conditioned specimens.

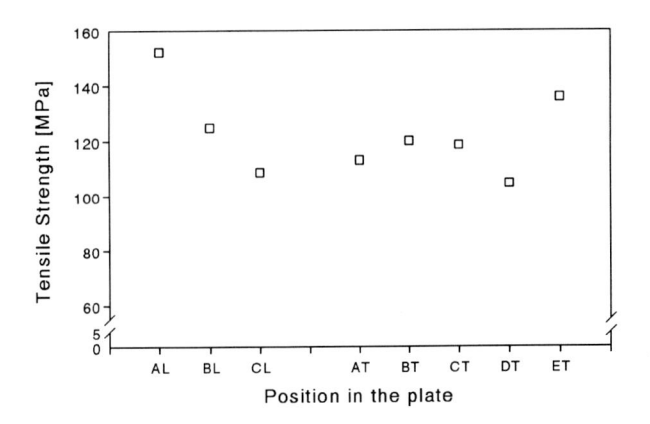

Chapter 3, Mechanical behaviour of composites 31

Figure 3.14 Variations of tensile strength over the plate, 5.75 mm conditioned specimens.

disappeared entirely. That this effect is much stronger for the 5.75mm plates compared to the 2mm ones will be no surprise if the thickness of the core layer, shown in Figure 3.15, is considered: The core layer in the thick plate accounts for approximately 50% of the total thickness, while this figure is 20% for the thin specimens. The indications of fibre orientation and thickness of layers are based on micrographic observations of polished sections. Thus the changes in the core over the width of the plate, result in higher differences of strength and stiffness in the case of the thicker specimen. The orientation in the shell layer is fairly constant over the width of the plate, and is aligned with MFD.

The opposite counts when the properties of specimens cut perpendicular to MFD are viewed: Variations with distance to gate (AT,BT,CT etc.) are relatively small. However, the strengths in Longitudinal and Transverse direction of the 5.75 mm plate (CL and CT) are approximately equal, while for the 2mm plate the strength in the perpendicular direction is considerably lower than in the longitudinal (MFD) direction. This again is caused by the differences in relative thickness of the core layer, shown in Figure 3.15. As the thickness of core and shell layers are approximately the same for the thick specimens, and the fibre orientation is respectively perpendicular to MFD and parallel to MFD in these layers, the properties in both directions are

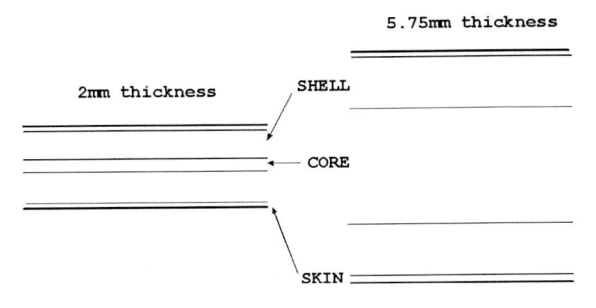

Figure 3.15 Comparison of relative thicknesses of core and Shell layers in 2 and 5.75 mm thick plates.

approximately the same. The relative increase in strength perpendicular to MFD near the end of the flow path, compare the strengths of DT and ET type of 5.75 mm thickness, is caused by the fibre orientation in the Shell layers. Fibres are not orientated by shear, and therefore average fibre orientation is much more perpendicular to the flow direction.

Thus the degree of anisotropy increases with the difference in core and shell layer thickness, and is generally higher in a thin specimen. On the other hand the degree of inhomogeneity of properties increases with increasing relative thickness of the core layer, and is generally higher in thick specimens.

The variations of tensile Modulus were very much comparable to those of tensile strength.

3.5.2 Influence of humidity

PolyAmide is very sensitive to moisture: Water is absorbed, which is bonded at the hydrogen bonds between the PolyAmide molecules. This influences strongly the mechanical properties, also for glassfibre reinforced PolyAmides [Akay,94]. In air of 23°C and 50% RH the water content will reach an equilibrium of approximately 2.5%. Immersed in water the material can absorb as much as 8% water. In glassfibre reinforced PolyAmide the problem is even more complex; the interface between fibre and matrix is weakened by the water. While the decrease in properties of pure PolyAmide recover as the water is evaporated from the material, a permanent reduction of properties remains in glass bead reinforced PolyAmide [Hartingsveldt,87]. Most probably in the case of GFPA this will be the same.

The material under investigation will normally not be used in a dry environment, reason why most experiments have been executed using conditioned material. As a comparison some experiments have been done with both dry-as-moulded and wet material, so as to know the properties for the extreme cases.

The tensile strength was measured for the conditions mentioned. In Fig. 3.16 the results are shown for different specimen types (see Fig. 3.12) cut from the 5.75mm thick plate. The influence of conditioning is similar for different specimen types, which can be seen more clearly

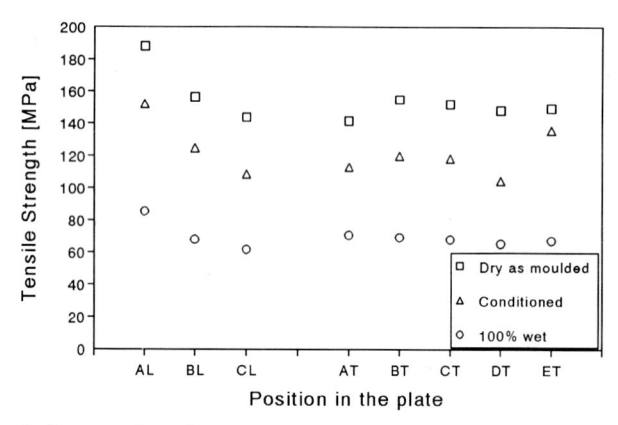

Figure 3.16   *Influence of conditioning on tensile strength. Overview over different specimen types of 5.75mm thickness.*

Chapter 3, Mechanical behaviour of composites 33

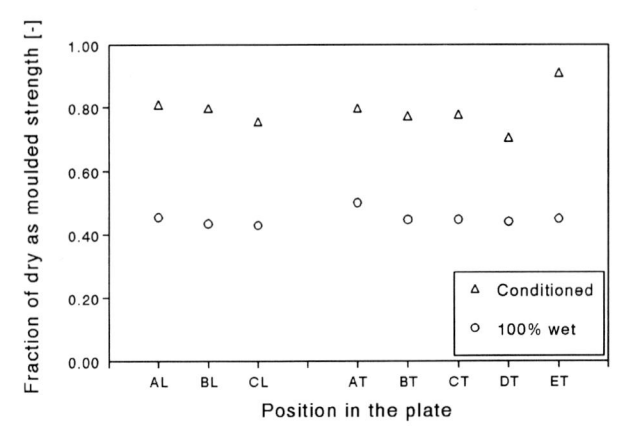

Figure 3.17   Comparison of decrease in tensile strength due to the water absorption, relative to the tensile strength for dry as moulded specimens.

in Fig. 3.17. Here the reduction of strength with conditioning is shown, with the dry as moulded material as a reference. It can be seen that, irrespective of fibre orientation, the relative decrease in strength due to the water absorption is approximately equal for all specimen types. The conditioned specimens have 80%, and the 100% wet specimens 45% of the strength of the dry as moulded material. Sole exception is the ET conditioned specimen, which shows a relatively high strength. No explanation can be given for this.

3.5.3 Influence of strain rate

Most plastics are visco-elastic materials, showing time and temperature dependent behaviour. Generally with increasing strain rate (thus shorter times) the strength increases, and the strain to fracture decreases. That fibre reinforced grades also show time dependent behaviour, can be seen in Fig. 3.18. Here the UTS is shown as a function of strain rate.

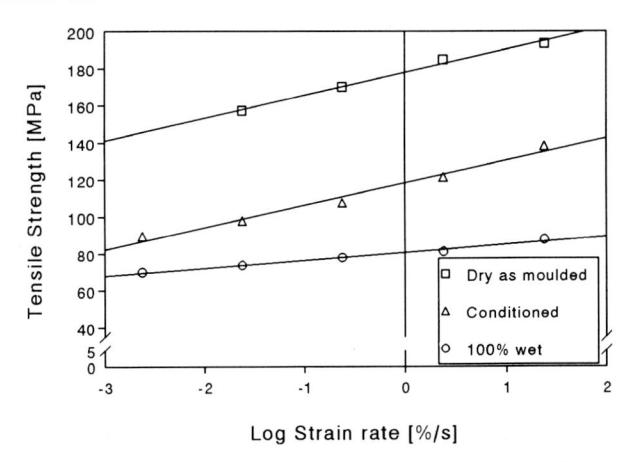

Figure 3.18   Influence of strain rate on strength. Effect of water absorption.

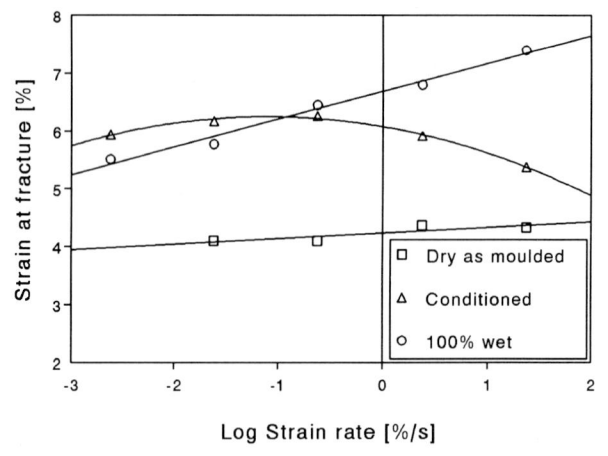

Figure 3.19   Influence of strain rate on strain at fracture. Effect of water absorption.

Both the dry as moulded as well as the conditioned material behave as expected, the strength increases with increasing strain rate. For wet material though, the strength is almost independent of strain rate. When looking at the strain at fracture, Fig. 3.19, the situation is not so clear. Depending on the conditioning, the strain at fracture can be independent of strain rate, or either increase or decrease. This shows that the effect of the visco-elastic behaviour of the matrix, can not be easily translated into a visco-elastic behaviour of the short fibre reinforced composite.

Firstly the effect of water on the fibre - matrix interfacial strength must be accounted for. This strength decreases with increasing water absorption. Thus the debonding of the fibres will begin at lower strains with increasing water content, causing the specimen to fail at a relatively low strain. This is the explanation for the relatively low strain at fracture of the wet specimens (Fig. 3.19), which is comparable to the strain at fracture for the conditioned specimens.
Secondly the increasing strain rate increases the notch sensitivity of the material, thus a crack that initiates in the specimen will grow faster. Thus the rest of the specimen is not strained to its full potential, and the specimen will fail at a lower strain.
The strain at fracture behaviour of the material as shown in Fig. 3.19 is the result of these competing effects.

3.6 Conclusions

The mechanical properties of a short fibre composite with a given fibre orientation can be predicted relatively accurate. Accuracy in the case of stiffness predictions is generally better than for strength. Mostly for not aligned fibres an averaging procedure is used, in which fractions of the fibres in different orientations are averaged over all possible angles. Per fraction the properties are assumed to be equal to that of an aligned composite. Thus the three different mechanisms are assumed to be independent, which is highly questionable especially in the case of strength. Furthermore at corners and stress concentrations already mentioned in section 2.6 and shown in Figures 2.7 and 2.8, calculation of properties from fibre orientation will need further research, as all available literature is on simple, plate geometries.

Micromechanical models in which the stiffnesses and strengths of fibres, matrix and interface are used to calculate product-properties do not give as good results as models in which longitudinal and perpendicular properties measured on the composite itself (with aligned fibres) are used. Especially the prediction of strength is a complex problem, which can not be solved easily using micromechanical modelling. However, micromechanical modelling can be very helpful in understanding what is the nature of the reinforcing effect of the fibres, and what is the relative influence of different factors on this.

The anisotropy and inhomogeneity is sometimes a benefit to the designer. However this will be only the case if the designer understands the mechanisms of fibre orientation and can adapt his design to the resulting fibre orientation, or influence the final orientation in the design by placement of the runners, wall thicknesses etc. We can conclude here that a thin specimen will have higher anisotropy compared to a thick specimen, as the greater part of the thickness consists of the highly aligned *shell layers*. However the inhomogeneity of properties (measured in one direction, on different locations) in a plate is less in the thin plate, as these inhomogeneities are mainly caused by fibre-orientation variations in the *core layer*, while orientation in the shell layers is fairly well aligned with MFD.

The influence of water absorption on the strength is substantial, with a maximum strength decrease of 55% when the material is saturated with water. The influence of water absorption on tensile strength is independent of fibre orientation: Specimens with different fibre orientation show the same *relative* decrease in strength. The influence of strain rate on tensile strength is as expected for visco elastic materials. For the strain at fracture the behaviour with varying strain rates is very much dependent on the conditioning of the specimens. Differences in failure mode of the interface may lead to results that differ from classic visco-elasticity.

4. Fatigue lifetime measurements

4.1 Introduction

The main purpose of the research is to investigate the relation between fibre orientation and fatigue behaviour. In this chapter the results for fatigue lifetime experiments are presented, on which the emphasis lies in this research. In Chapter 5 results of fatigue crack propagation experiments are given.

A lot of research has been published on fatigue lifetime experiments. The majority of these were written with an interest in the material performance only, and standard injection moulded test specimens were used. This approach ignores the influence that fibre orientation has on the fatigue behaviour. In some articles the fibre orientation has been given some consideration, though this is mainly in fatigue crack propagation (see Chapter 5).

In this chapter measurements on fatigue lifetime of glassfibre reinforced PA6 will be presented. The influence of fibre orientation on fatigue lifetime is accounted for by taking specimens containing different orientation distributions, resulting in different tensile properties for these specimens, e.g. different strengths and stiffnesses. Apart from the influence of *fibre orientation* on fatigue, also the influence of *fatigue frequency*, *specimen thickness*, *material humidity* and *fibre-matrix bond quality* has been investigated.

4.2 Experimental procedure and materials

The material used was PolyAmide 6 containing 30%Wt. of glassfibres; Akulon K224-G6 provided by DSM, the Netherlands. Square plates of 100x100 mm and of 2 and 5.75 mm thickness, and square plates of 90x90mm with thicknesses of 2 and 6mm were injection moulded from this. (See Appendix I for the injection moulding conditions) The lifetime experiments were executed on dog-bone type specimens that were cut from the plates, according to Fig. 3.12. This was done using a Roland PNC-3000 Computer Aided Modeling Machine. The fatigue experiments were executed in a servo-hydraulic MTS type 810 bench under load control in *tension*, at a frequency of 1 Hz unless otherwise stated. The minimum to maximum load ratio R was 0.1. Temperature during testing was 23°C ± 1 and relative humidity 50% ± 5. During the experiments the temperature at the surface of the specimens was measured using an infrared contactless thermometer. Most specimens were conditioned by exposing them to laboratory air for about 1 year. Also specimens that were dry as moulded were used, as well as completely wet specimens, which were emerged in water for a minimum of three weeks to ensure a high water content. The different specimens are referred to as conditioned, dry as moulded and 100% wet respectively. For the testing of these specimens special conditions were created: The dry as moulded specimens were tested in an atmosphere with low humidity, of 30% ± 5 RH. Weighing of the specimens before and after the experiment showed that no significant amount of water was absorbed during the test. The wet specimens had to be fatigued *in water* to ensure that no water would evaporate, in case of low loads / long lifetime tests.

During the test the load, elongation and surface temperature were continuously monitored using a computerised system. This enabled the calculation of cyclic Modulus, creep during fatigue and the energy dissipation. The energy dissipation was calculated from the hysteresis loop every 5 cycles. The result of each test thus is much more than just the fatigue lifetime, a test sheet is produced where elongation, creep speed, cyclic Modulus, energy dissipation and surface temperature during the test are given, see Appendix II.

4.3 Results

4.3.1 S-N curves

The results of lifetime experiments, where a specimen is subjected to an oscillating load until failure, can be represented in various ways, however the most common is the S-N curve or Wöhler curve [Hertzberg,80]. The fatigue stress amplitude is normally plotted on the vertical axis, while the logarithm of the number of cycles to fracture (the lifetime) is put on the horizontal axis. As the maximum stress is close to twice the stress amplitude, and the R-value is constant in all experiments (R=0.1), in this case on the vertical axis the maximum <u>fatigue stress</u> is plotted. In Fig. 4.1 an example is shown. For three cases results are given, the symbols indicate fracture of the specimen at a certain fatigue stress, thus indicating the fatigue strength of the specimen for that lifetime. Each symbol is an average of 10 experiments in this case. In later experiments always a minimum of 5 experiments per load level was executed. The striking feature is that the curve for non-reinforced PolyAmide crosses the curve for the CL type specimen. At long lifetimes the not-reinforced grade gives better fatigue performance than the reinforced grade!

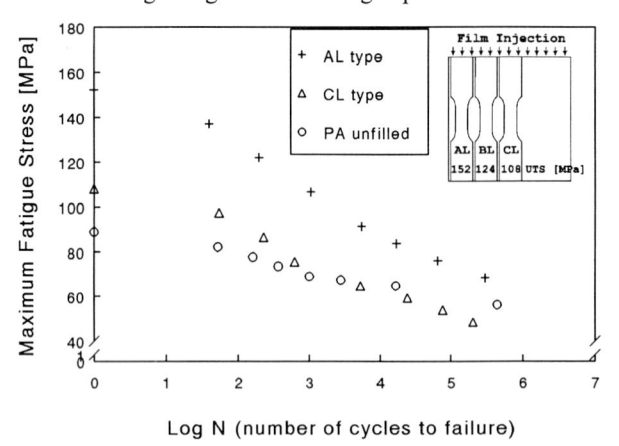

Figure 4.1 *Example of a S-N or Wöhler curve. Lifetimes are given for AL and CL specimen types. For comparison the curve for not reinforced PolyAmide is given as well. On the vertical axis (log(N) = 0) the tensile strength is indicated.*

4.3.2 Effect of frequency

The effect of frequency on fatigue is complex. On the one hand an increase in frequency results in an increase in loading rate. As the material, like all polymers, is a visco-elastic material, the properties are dependent on the loading rate. At a higher loading rate the material has higher strength, see § 3.5.3. Therefore the fatigue lifetime (in cycles) will increase at a higher frequency [Remmerswaal,90]. At low frequencies the fatigue can become creep-dominated, leading to a decrease in number of cycles to failure. On the other hand the energy dissipation due to the cyclic deformation, will heat up the material. An increase in frequency will result in a higher temperature of the specimen. The higher temperature results in a decrease of properties, and a decrease in fatigue lifetime. Temperature increases at a crack tip have been reported to be high, up to 90°C [Wyzgoski,90] for *not-reinforced PA*, and even an increase of 30°C at 0.5 Hz was reported. For reinforced material [Wyzgoski,91] the temperature increases are far less, and very little influence on fatigue crack propagation was found. The influence of reinforcement is a decrease in cyclic deformation, so the crack tip temperature rise decreases with an increasing fibre content [Lang,83]. Suzuki and Kunio [Suzuki,88] found a linear relation between temperature increase (in lifetime specimens without crack) and frequency.

As the influence of frequency is composed of these competing effects, and the effect of frequency is lower with increasing fibre content, an evaluation of fatigue behaviour with frequency is necessary. Furthermore the viscoelastic behaviour of nylon changes with the water content [Lang,83], which influences the heat build-up as well as the sensibility to the loading rate, §3.5.3.

<u>Effect of frequency on fatigue lifetime</u>
To evaluate the overall effect of frequency on fatigue lifetime, experiments were done at various frequencies, for conditioned specimens. This in order to decide at what frequency future experiments should be executed. Frequencies of 2, 1, 0.5 and 0.2 Hz were compared, Fig. 4.2. At 1 Hz lifetime is highest, indicating that at higher frequencies the energy dissipation causes a decrease in lifetime, while at lower frequencies creep begins to play a role, though the effect is smaller than that of temperature increase.

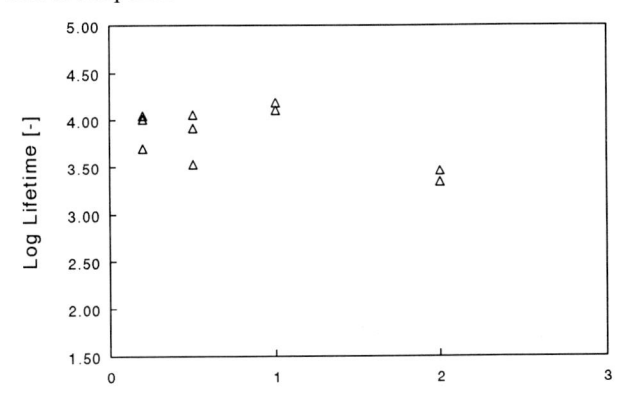

Figure 4.2   *Lifetimes for BL type specimens of 5.75 mm thickness, fatigued at 50% of UTS, and various frequencies.*

The effect of temperature increase of the specimens is stronger for thick specimens, as these can not transfer heat to the environment as easy as thin specimens. The effect of frequency on fatigue lifetime therefore is dependent on the thickness of the specimens. Furthermore the effect for the material, Polyamide 6, is stronger than for most other thermoplastics. As the glass transition temperature T_g for PA6 in dry condition is approx. 60 °C, and is lowered by the absorption of water to 25 - 35 °C for normal conditioning. Internal damping of thermoplastics is very high near T_g, and therefore internal damping for conditioned PA6 at room temperature is high.

From Fig. 4.3 it can be concluded that for thin specimens a decrease of lifetime at high frequencies up to 10 Hz does not occur. On the contrary the fatigue lifetime shows an increase with increasing frequency. This is the effect of the increased strength and stiffness at higher loading rates. Variations in lifetime with frequency are less pronounced than for the thick specimens, except at high fatigue loads.

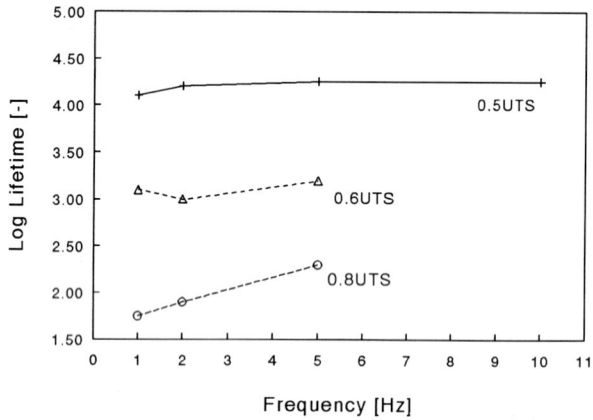

Figure 4.3   *Lifetimes for EL type specimens of 2 mm thickness. Fatigued at various frequencies, and at fatigue stresses of 0.8UTS, 0.6UTS and 0.5UTS.*

From the measurements shown here it was concluded to do future fatigue tests at a frequency of 1 Hz, because of the temperature effect at higher frequencies, and the creep effect at lower frequencies. A frequency of 1 Hz was used for all experiments, including those using thin specimens. This to avoid differences between specimens of different thickness, in order to compare these more easily. Another reason to choose this frequency is the duration of experiments, at 1 Hz, this is still acceptable (up to 10^6 cycles).

<u>Contribution of creep to fatigue</u>
To assess if creep (in the loaded part of the cycle) influences fatigue, a series of experiments was executed with a constant load, and the creep was measured. EL type specimens of 2 mm thickness were used. In Fig. 4.4 the creep speed is shown for these experiments, at different fractions of tensile strength. The creep speed at constant load is the increase in elongation per second, during secondary creep, where the creep speed is constant. This is analogous to the situation in cyclic creep, as will be discussed in 4.3.3.

Chapter 4, Fatigue lifetime measurements   41

As can be seen in Fig. 4.4 the logarithm of creep speed in tensile creep is linear with the load. For comparison the creep speed for cyclic (fatigue) experiments is also shown, for frequencies of 1-5 Hz. The creep speed in fatigue at the load shown is at least a factor 100 more than that in tensile creep, at the same load. Note here that in fatigue the maximum load is applied only a fraction of the total loading time. Thus the constant-load creep in fatigue influences only a fraction of the load cycle.

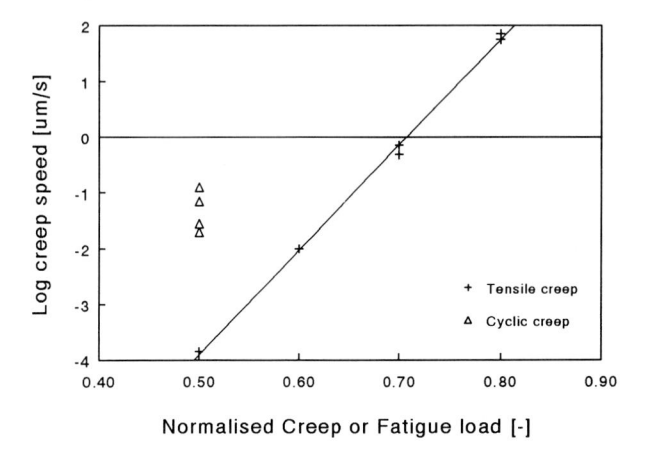

Figure 4.4   *Creep speed (elongation increase per second) in fatigue respectively tensile creep experiments. On the horizontal axis the fraction of tensile strength at which the creep or the fatigue test was executed.*

It can be concluded therefore that the contribution of tensile (constant force) creep is negligible in fatigue at low loads. The oscillating load in fatigue brings about much more creep than a constant load.

Temperature increase at the surface
The temperature measurements show that for *all* experiments the temperature of the specimens does increase, even at this relatively low frequency of 1 Hz.
For low loads (up to 0.6UTS) normally equilibrium is reached, the temperature increase at the surface will be less then 5°C. At higher loads the temperature at the surface of the specimen does not reach equilibrium, but increases linearly with time, until failure. The temperature increase can be as high as 10°C. Naturally, the temperature inside the specimen will be higher than the temperature at the surface.

It is impossible to execute the fatigue experiments at a frequency low enough to prevent any temperature increase. However the temperature increase measured is relatively low, especially at the low loads that are of use for practical long livety design purposes.

4.3.3   Creep Speed method

When the possibility to execute experiments for measuring the development of damage during fatigue was evaluated, it was realised that a method should be devised to account for the scatter in lifetime. If a fatigue experiment is stopped, for the determination of the damage caused by fatigue, it is obvious that the fatigue experiment can not be conducted to the end. As the scatter in lifetimes for this material was found to be approximately a factor of 2 between minimum and maximum lifetime at a certain load level, the percentage of lifetime at which the experiment was stopped is not known exactly. An estimate can be made using the average lifetime at that load level, but still the actual lifetime of the experiment can vary by 50%. To overcome this problem, various parameters that can be measured during fatigue were correlated with fatigue lifetime. The creep speed turned out to be the only reliable parameter.

The creep speed V_c as used here is identified as the increase in elongation at maximum load per cycle, see Fig. 4.5. In secondary creep the creep speed is practically constant, and therefore is not sensitive to the number of cycles over which it was measured. In practice for specimens that were fatigued up to fracture, V_c was calculated from 20% to 80% of lifetime.

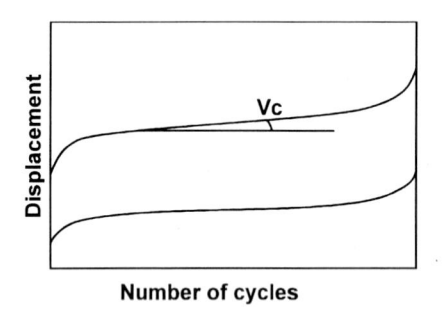

Figure 4.5   *Calculation of the creep speed from the fatigue creep curves. The top curve is the elongation at maximum load, the bottom one is at minimum load. Creep for both is approximately equal. From the difference between both curves the Elastic Modulus can be calculated.*

In literature little has been reported about creep in fatigue. Creep curves have been reported by Jinen [Jinen,86 and 89], for the case of carbon fibre reinforced PA6. This in order to compare the relation between the damage to the material by the fatigue process, and that arising from creep. From the creep results it was concluded that the effect of the repeat effect in fatigue on the material deformation is about twice as large as the effect of static creep. A relatively high influence of static creep compared with the measurements presented in Fig. 4.4, which can be explained in part by the low frequency used, of 0.1 - 0.2 Hz.

In the investigation presented here however, the creep is used in first instance as a tool for predicting the lifetime of a specimen, before failure.

Creep speed (V_c) during secondary creep gives a correlation with lifetime. Fig. 4.6 shows an example of the correlation between log V_c and log N, at one load level and for one specimen type. All variation in lifetime is due to natural variation between specimens. Using the creep speed to calculate the expected lifetime, the uncertainty about the lifetime decreases. The decrease is by a factor of 2 - 3, compared to using the mean value of the lifetime determined from experiments. This means that the percentage of the fatigue lifetime at which the experiment was stopped is known more exactly. See for more examples Chapter 6.

Chapter 4, Fatigue lifetime measurements

The fact that V_c correlates with N indicates that the variation in the lifetime of the specimens is determined by differences in the specimens that are of *global character* (aspects of the specimen as a whole), as creep is a *global parameter*. Consequently *local variations* (e.g.. scratches,

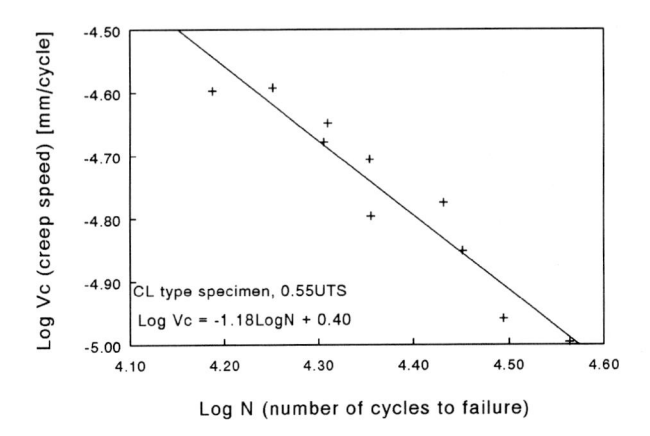

Figure 4.6 Log V_c *(creep speed) - Log N (lifetime) relation for one specimen type (CL, 5.75 mm thick) at one load level (0.55UTS).*

notches or voids) have a small influence on fatigue lifetime. This can be explained by the fibres. Contrary to unfilled polymers, SFRTPs are full of impurities (the fibres) and notches (the fibre ends). Notch effects of scratches at the surface or otherwise, therefore are of minor importance. The collective notches (fibre ends), may have an effect though on both the creep and the fatigue behaviour, which explains the correlation.

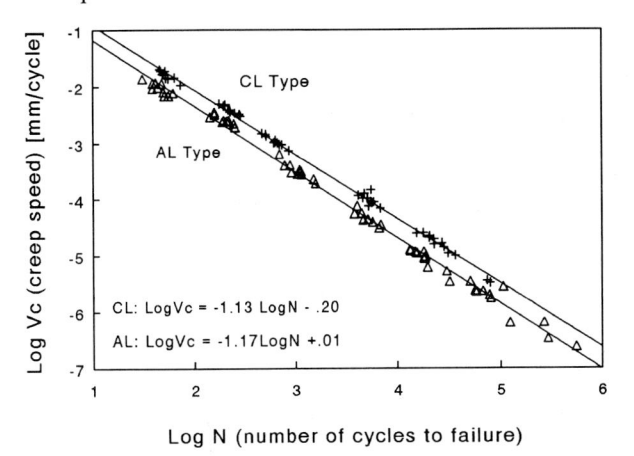

Figure 4.7 Log V_c *(creep speed) - Log N (lifetime) relation for AL and CL type 5.75 mm thick specimens.*

The degree of correlation between Log V_c and Log N at one load level (R-squared), is not constant for the whole load range. The correlation is best for conditioned specimens and at relatively high load levels (0.7-0.8UTS). For higher load levels the experiments are too short to make an accurate measurement of V_c. R-squared of the correlation is usually above 0.75. Correlation is less for dry as moulded specimens. This can be explained (see Chapter 8) by the fact that less plastic deformation occurs in these dry specimens. This increases the notch sensitivity of the material and therefore the importance of local variations on the fatigue lifetime.

The relationship between V_c and N holds not just for one load level, but also for the whole load range tested, see Fig. 4.7. In this graph it is visible that the CL type specimens possess the higher creep speed at a given number of cycles to failure. Creep speed is approximately a factor 2 higher as compared to the AL type specimens. This corresponds with the 2-times higher fracture strain of the CL type specimen in a tensile test due to the higher thickness of the core layer and therefore less fibres aligned in loading direction. BL type specimens (with strength and stiffness in between AL and CL specimens) have a creep speed in between. This is not shown for clarity. This correlation over the whole load range indicates that no change in failure mechanism takes place over the load range examined, which would probably induce a change in the creep behaviour as well. SEM fractography (Chapter 7) supports this theory: On all fracture surfaces of fatigued specimens a microductile and a microbrittle area can be seen. The size of the microductile area increases monotonically with decreasing fatigue load and increasing lifetime, while the appearance of the microductile and microbrittle areas do not change.

4.3.4 Modulus and Energy dissipation

The cyclic Modulus is calculated from the displacements and loads, as shown in Figure 4.8, by dividing the stress amplitude by the strain amplitude.

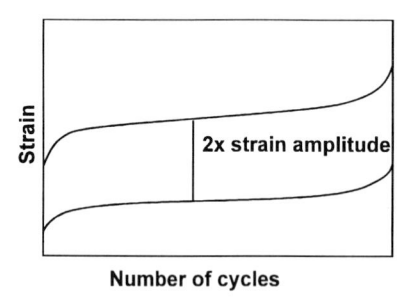

Figure 4.8   Calculation of cyclic Modulus.

The cyclic Modulus decreases linearly with number of cycles during the experiment, in some cases complemented by a stronger decrease towards the end of the experiment. Roughly the decrease is 20% for loads of 0.8 - 0.7UTS, while for loads lower than 0.6UTS it is 10%. Part of this decrease in Elastic Modulus can be explained by the temperature increase of the specimen during the test.

The energy dissipation can be calculated from the hysteresis loop. In Fig. 4.9 the development of these hysteresis loops is seen, during one experiment.

Chapter 4, Fatigue lifetime measurements 45

The change in the hysteresis loops at various percentages of lifetime can be seen: The main axis of the hysteresis loop has a decreasing angle towards the X-axis, indicating the decreasing cyclic Modulus. The area of the hysteresis loop increases, showing the increasing energy dissipation. The position of the hysteresis loop shifts to higher displacements, this is the cyclic creep effect.

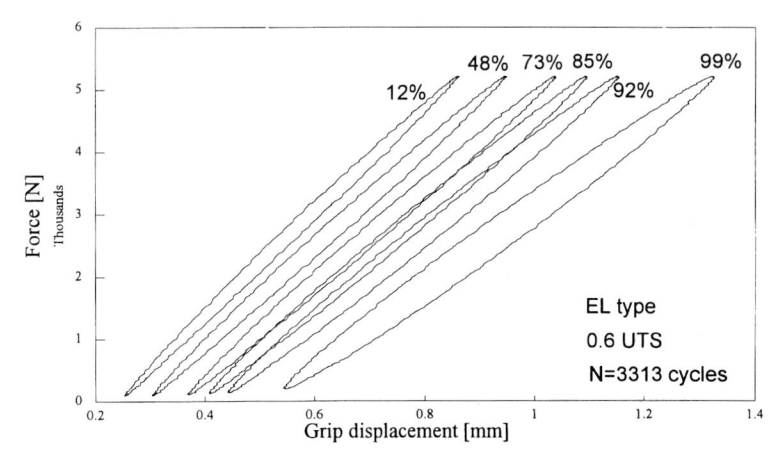

Figure 4.9   *Development of hysteresis loops during the experiment. Percentages indicate the percentage of lifetime at which the hysteresis loop was made. Changes in cyclic Modulus and creep are also visible.*

It can be noted here that the creep (change in position of the hysteresis loop) is most easily seen in the graph, it is the property that changes most during the experiment.

The energy dissipation increases linearly during the experiment, resulting in a total increase of roughly 50%. The energy dissipation per cycle ranges from 100 mJ *per cycle* for the experiments at low loads to 500mJ for the higher loads. The energy dissipation for different specimen types is the same if the experiments are executed at the same normalised load.

4.4 Effect of orientation, normalised S-N curves

In literature a method was found for normalising S-N-curves [Adkins,88, Mandell,80, Dally,69]. It was found there to be a good method to compare fatigue-behaviour of different materials, as the fatigue strength is compared to the tensile strength. Practically this is done by making a S-N or Wöhler curve, with on the vertical axis not the fatigue stress (or stress amplitude) but the fatigue stress divided by the tensile strength (UTS).

For the current investigations, where the effect of fibre orientation on fatigue behaviour is investigated, it was thought to be a useful method to compare fatigue behaviour of different specimen types, which show different tensile strengths because of different fibre orientation. In Fig. 4.10 the normalisation procedure is shown for two specimen types, of 5.75 mm thickness. As can be seen in Fig. 4.10 the curves for the two specimen types, coincide after normalisation! Thus normalisation of the S-N curve leads to a "Master Curve" which is independent of fibre

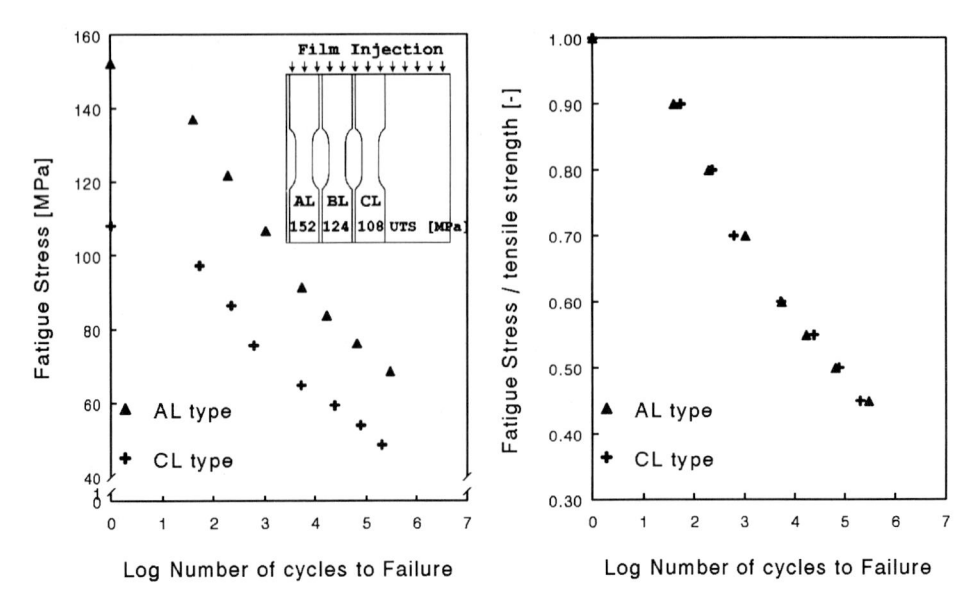

Figure 4.10   The normalisation procedure: On the left the S-N curve, on the right the normalised S-N curve. On the Y-axis the tensile strength is given, for these 5.75 mm thick specimens.

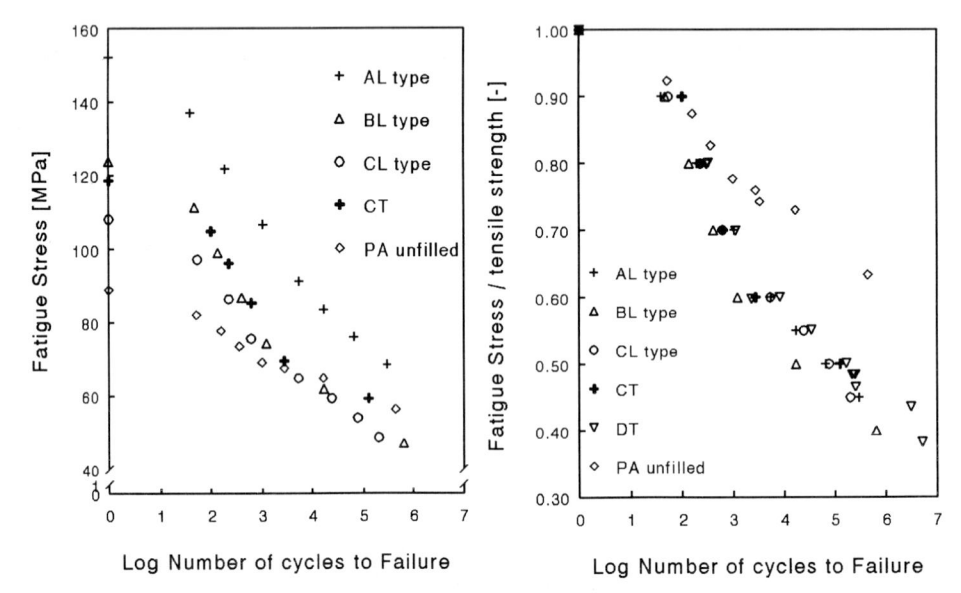

Figure 4.11   Fatigue results for various specimen types, conditioned, 5.75 mm thickness. On the left the S-N curve, on the right the normalised S-N curve. Thickness of the pure PA specimen is 4 mm.

orientation. In Fig. 4.11 it is shown that this normalisation procedure is applicable for different specimen types, all conditioned specimens of 5.75 mm thickness. A factor which is of importance is the value for UTS that is used. As was shown in § 3.5.3 the tensile strength is dependent on the strain rate. For this investigation the strain rate that is used was 2.3 %/s, corresponding with a crosshead displacement of 50 mm/min. For the conditions of the tensile experiment to be the same as for the fatigue experiment, the strain rates in both cases should be comparable. The tensile experiment can be seen as half a cycle of a fatigue experiment, with the specimen failing in the first load cycle. As the frequency used for the fatigue experiments was 1 Hz, the specimen should fail in approximately 0.5 seconds in the tensile test. From Figure 3.19 the time to failure in the tensile tests can be calculated. For the crosshead displacement of 50 mm/min this is 1.7-3 seconds, while for the tests executed at 500mm/min the time to failure is between 0.18 and 0.3 seconds. The conclusion is that the strain rate chosen is slightly to low, and that the tensile strengths resulting from the faster tensile tests might give a better result. In Fig. 3.18 the difference in UTS between both strain rates can be seen. If a slightly higher value for UTS would be used in the normalisation procedure, the tensile strength (shown on the Y-axis) would be on one line with the fatigue results. However, the effect of using a different UTS will have little or no effect on the normalised fatigue results from the different specimen *types*. Therefore the UTS measured at a crosshead speed of 50 mm/min is used throughout the research here presented.

Generally all experiments fall on the one master curve, except for the BL type specimen, which shows a lower fatigue strength. Explanation for this lower value can be found in the fractured specimens: The fracture in this case is not perpendicular to the specimen axis, but at an angle of approximately 45°. This due to the fibre orientation at the location of the BL type specimens, which can be seen in Fig. 1.1. The orientation in the core layer is at an angle of approximately 45° to the specimen axis. Thus a shear mechanism occurs, since the highest shear stresses are at 45°. The fatigue results show that this shear mechanism affects the fatigue results more than the tensile strength. In Chapter 8 a possible cause for this will be discussed.
For comparison also the values for not-reinforced PolyAmide are included in the graph. As was already shown in Fig. 4.1 the fatigue strength of unreinforced PA can be higher than the fatigue strength of the reinforced grade, if this has unfavourable fibre orientation. In Fig. 4.11 it can be seen that if the normalised fatigue strength is used, the reinforced grade shows worse fatigue behaviour compared to the unreinforced PolyAmide, for all types of specimen, even for aligned fibres.

That the existence of the master curve is not just present for the *conditioned* specimens can be seen in Fig. 4.12, where the normalised fatigue strengths for *dry as moulded* material are shown. A coinciding of the curves for the different thicknesses is also present, as for the conditioned material. In Fig. 4.13 the normalised S-N curve is shown for *100% wet* specimens of 2 mm thickness.
It is shown that the fatigue behaviour of specimens with different fibre orientation distributions can be represented by one Master Curve or normalised S-N curve. In this way the fatigue behaviour is related to the tensile strength. This normalisation procedure leads to master curves for conditioned, dry as moulded as well as wet material. However, the Master Curves for these different conditions are not the same. The position of the Master Curve is affected by material humidity, as well as fibre - matrix interfacial bond quality and fibre length, see §4.5 and §4.6.

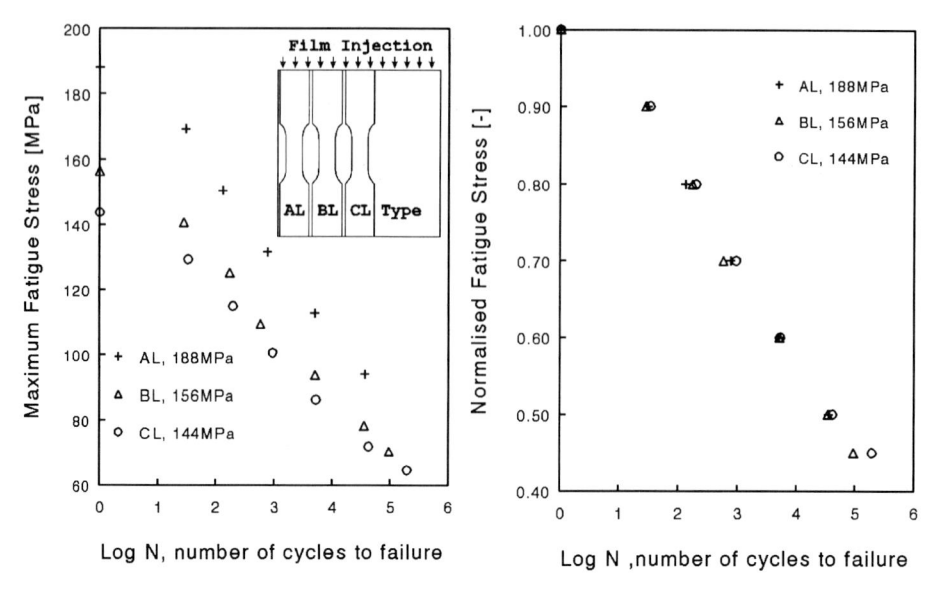

Figure 4.12 Fatigue results for dry as moulded specimens, 5.75 mm thickness. On the left the S-N curve, on the right the normalised S-N curve.

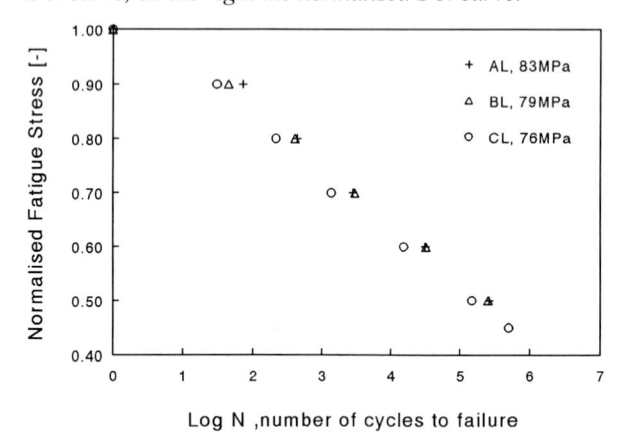

Figure 4.13 Normalised S-N curves for 100 % wet, 2 mm thick specimens.

4.5   Effect of humidity

The effect of humidity on glassfibre reinforced material is two-fold. On the one hand the amorphous regions in the PolyAmide absorb water, which is bonded at the hydrogen-bonds between molecules. The distance between PolyAmide molecules increases, increasing volume and increasing chain mobility. This results in a lower elastic Modulus and strength, and a higher ductility. On the other hand the water affects the fibre-matrix bond quality negatively [Hartingsveldt,87].

Chapter 4, Fatigue lifetime measurements 49

The effect of humidity on fatigue is shown in Fig. 4.14 for 2 mm and 5.75 mm thick specimens. Results are contradictory: In the 2 mm thick specimens a clear influence of humidity is seen, while for the thick specimens no difference can be noted. The graphs shown use the normalised fatigue stress; thus for the 5.75 mm thick specimens the effect of humidity is the same for the Fatigue Strength and for the Tensile Strength.

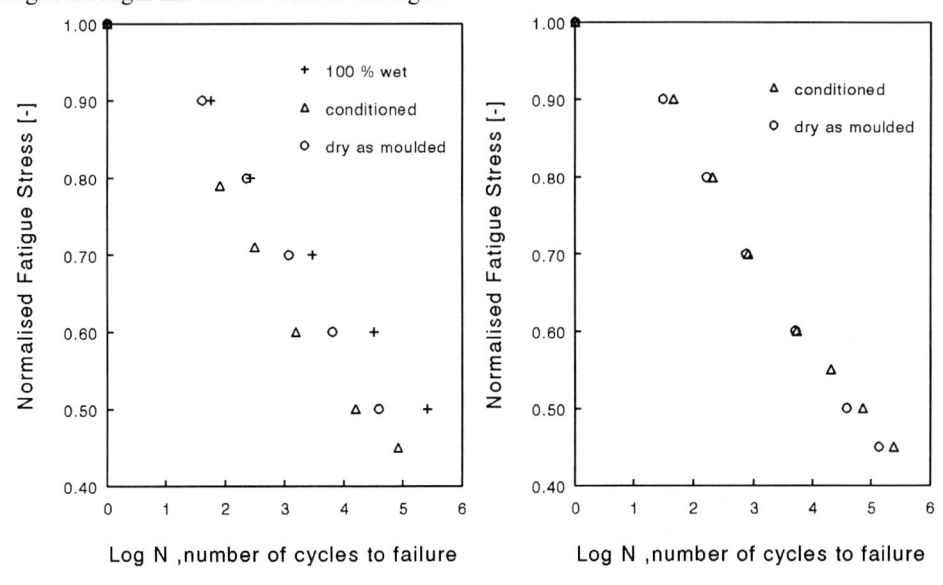

Figure 4.14   Normalised S-N curves for specimens with different humidity. Left: 2 mm thickness, right: 5.75 mm thickness.

For the 2 mm thick specimens the effect of humidity is complex, from dry to conditioned material the fatigue performance is worse, while the fatigue performance for 100% wet material is even better than for dry material. This is caused by an effect which will be discussed more in detail in chapters 7 and 8; the different role of the fibre - matrix interface in fatigue and in tensile experiments. In tensile experiments the interfacial strength for dry as moulded and conditioned material is sufficiently high for the interface not to fail. The matrix fails, close to the fibre. However, if the interfacial strength is lowered enough, in 100% wet material, the interfacial failure site shifts from the matrix, to the interface itself. On the contrary in fatigue the interface is always the location of failure: Fatigue affects the interfacial strength. Thus, when a little water is absorbed (from dry as moulded to conditioned), the interfacial bond strength is lowered, which has an effect only on the fatigue strength. If more water is absorbed, the interfacial strength is lowered enough for the interface itself to fail in the tensile experiments as well. From the results it can be concluded that this causes the tensile strength to decrease relatively more than the fatigue strength. This because the normalised fatigue strength for 100% wet material shifts to higher values, compared to dry as moulded and conditioned material.

What the reason is for the difference between 2 and 5.75 mm thick specimens is unknown, the tensile strength for the AL type specimens of both 2 and 5.75 mm thickness is 188 MPa and 152 MPa for dry as moulded and conditioned state respectively. Because the tensile strength of the

2 and 5.75 mm AL type specimens is the same, the possibility that the thick specimens were not conditioned sufficiently can be excluded.

4.6 Effect of bonding and fibre length

In a series of experiments executed for DSM, the Netherlands, the influence of fibre - matrix bonding on fatigue was investigated. This was done by investigating two materials, the standard Akulon K224-G6 used in all experiments, and an experimental grade of the same material, difference being an improved fibre coating, which gives an improved bonding between fibres and matrix. Fatigue behaviour of both materials was compared. For both materials no difference in tensile strength was found.

Experiments were done on 5.75 mm and 2 mm thick material. For the first only dry as moulded material was used, while for the thin material also tests were done with conditioned and 100% wet specimens. In Fig. 4.15 the results for thick specimens are shown, while in Fig. 4.16 the results are shown for the 2 mm specimens.

In all cases the experimental material shows an improved fatigue strength, while the tensile strength is approximately the same, except for the case of the 2 mm wet specimens; the experimental grade has a strength of 87 MPa, while the strength for the standard grade is 83 MPa.

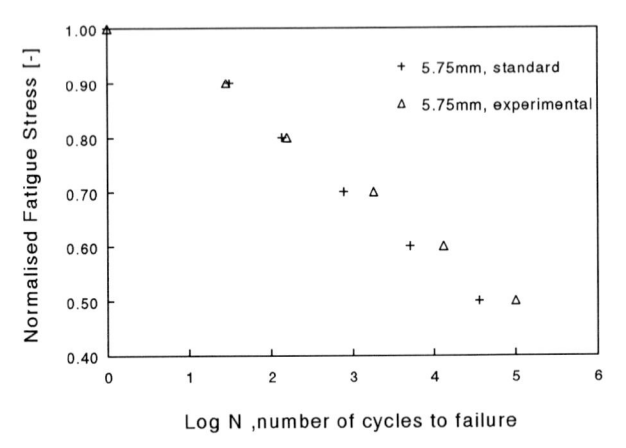

Figure 4.15   Lifetimes of standard and experimental GFPA. 5.75 mm thick, dry as moulded.

It was suspected that the different processing for both grades could induce a difference between both grades of material, affecting the fatigue behaviour. The experimental grade was compounded on a different extruder than the standard grade, which might affect the fibre length. Fibre lengths for the specimens were measured, results are summarised in Table 4.1. There are no significant differences in fibre length between the standard and the experimental grade: the differences in fatigue behaviour are therefore caused by the better fibre - matrix bonding of the experimental grade, because no significant other differences between both materials exist.

Chapter 4, Fatigue lifetime measurements

However, it can be seen in Table 4.1 that a difference in fibre length exists between specimens of different thickness. This of course can affect the fatigue behaviour, therefore in Fig. 4.17 the fatigue performance of thick and thin (dry as moulded) specimens is compared. No significant difference in tensile strength (UTS) was found.

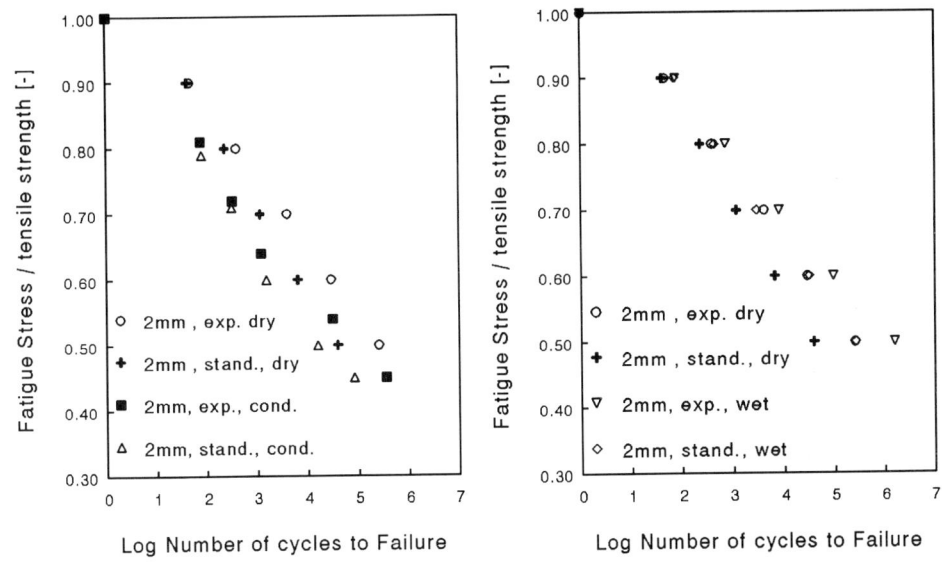

Figure 4.16   Comparison of standard and experimental material. 2 mm thickness, dry as moulded, conditioned and 100% wet.

thickness/ fibre	No. fibres measured	No. Average length	Length average	% fibres > 150 µm
2 mm, experimental	247	255 µm	315 µm	93.8 %
2 mm, standard	187	256 µm	307 µm	93.6 %
5.75 mm, experimental	244	204 µm	258 µm	88.5 %
5.75 mm, standard	211	207 µm	276 µm	82.9 %

Table 4.1   Comparison of fibre lengths in 2 and 5.75 mm thick mouldings. The Number of fibres measured are listed, the number avarage, the length average and the percentage of fibres which have a length over 150 µm.

An influence of thickness is not visible for the standard material, while for the experimental material the thin specimens, with the higher fibre length, show a small increase in fatigue lifetime. Of course other differences between both thicknesses play a role, not only the fibre length. For example the different thermal behaviour at different specimen thickness, the 2 mm specimens can transfer heat to the environment faster. This will affect the fatigue behaviour.

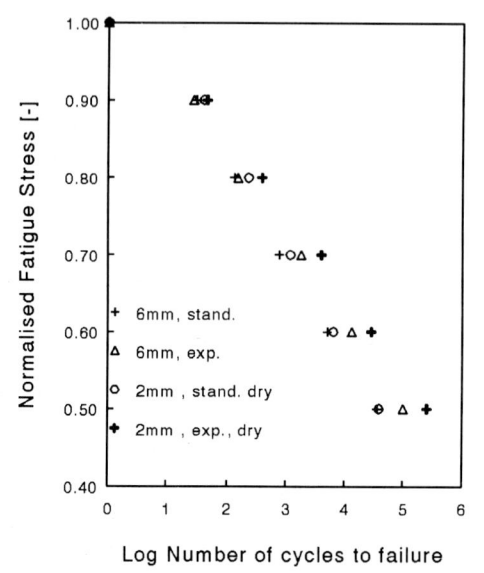

Figure 4.17   Thickness influence on fatigue behaviour. All dry as moulded specimens.

4.7   Conclusions

It is shown in this chapter that fibre orientation has a profound effect on the fatigue lifetime behaviour.

The effect of orientation on tensile strength and on fatigue strength is identical: If a change in orientation leads to a certain percentage increase in tensile strength, the percentage increase in fatigue strength is the same. In practice this means that a "Master Curve" can be used, a normalised S-N curve which is valid for specimens that differ in fibre orientation. Thus, when the Master Curve is known, measurement of the tensile strength of a specimen or product with a different fibre orientation suffices to know the fatigue behaviour. The normalisation procedure results in a Master Curve for conditioned specimens, but also for dry as moulded and wet specimens, although the Master Curves for these differently conditioned materials are not the same. Thus the procedure of normalisation of the fatigue stress by the tensile strength in order to obtain a Master Curve can be used for specimens with different fibre orientation only. Specimens that are different in other ways, like conditioning, fibre length or fibre - matrix bonding do not coincide to a Master Curve after normalisation.

In principle specimens of different thickness should be comparable as well. However the different mould can have an effect on fibre length, which in general will not be the same for both moulds, and may cause different fatigue behaviour. Also different thicknesses will cause different thermal behaviour (especially at higher frequencies), inducing differences in fatigue behaviour as well.

Apart from the effect of orientation on fatigue behaviour, other parameters have been investigated. Increases in fibre length as well as in fibre-matrix bond quality both favor fatigue performance. The effect of humidity is generally negative, both for tensile as well as fatigue properties. However, when looking at the normalised properties, the effect on fatigue can be favourable, when at high water contents the tensile strength is affected stronger than the fatigue strength. In Table 4.2 an overview of the influence the various parameters have on tensile strength, and absolute and relative fatigue performance is summarised.

Factor:	Tensile strength	Fatigue strength	Normalised Fatigue str.
Fibre orientation	+	+	none
Fibre length	+	+	+
Fibre coating	none *	+	+
Humidity	-	-	- or +

Table 4.2 An overview of the influence various factors have on tensile strength, fatigue strength, and normalised fatigue strength. *: above a certain minimum interfacial strength.

5. Fatigue Crack Propagation experiments

5.1 Introduction

A lot of research has been published on fatigue crack propagation (FCP) in short glassfibre reinforced thermoplastics. In the majority of these the fibre orientation was of no or little importance. For example Lang et al. in 1983 found no difference in FCP behaviour between transverse and longitudinal orientations. An observation that they do not make however, is that in their case the thicknesses of Core and Shell layers were similar, resulting in similar properties in both directions.

However, in a few articles the results were correlated with either the fibre orientation [Karger-Kocsis,90 and 88, Voss,88, Friedrich,86] or mechanical properties that are influenced by the fibre orientation, like the Elastic Modulus [Wyzgoski,91 and 88].

In the first case FCP is related to a microstructural efficiency factor M, which includes the relative thicknesses of core, shell and skin layers, the fibre orientation in these layers, the fibre length and the fibre volume fraction. Although varying orientation and layer thicknesses were found over the moulding, an average was used to calculate the microstructural efficiency factor for the entire crack length.

Wyzgoski and Novak relate the crack growth rate to the strain energy release rate (see equations 5.4 and 5.5) which is calculated using the stress intensity and the Elastic Modulus *in the loading direction*.

In one article an attempt is made to fit behaviour for different types of plastics with one single FCP curve [Chow,91]. da/dn is related to the strain energy release rate ΔG, and the difference between the maximum G and the critical G representing the fracture toughness. In this way the fatigue behaviour of PMMA, PVC and short fibre reinforced nylon 6.6 could be represented with one line.

In this Chapter, as in Chapter 4 for lifetime experiments, the fatigue crack growth behaviour will be related to tensile strength. It is investigated if the relation between fatigue stress and tensile strength found in lifetime measurements, can be applicable to fatigue crack propagation as well.

5.2 Experimental procedure and materials

The material used was PolyAmide 6 containing 30%Wt. of glassfibres, see §4.2. Square plates of 100x100 mm and of 2 and 5.75 mm thickness were used for fatigue crack propagation experiments. A notch of 16mm length was machined in the centre of the plate, perpendicular to the loading direction (Fig. 5.1). The notch tip radius was 0.01 mm. The fatigue experiments were executed in a servo-hydraulic MTS type 810 bench under load control in *tension*, at frequencies of 10 and 4 Hz. The minimum to maximum load ratio R was close to 0, a negligible minimum load was maintained throughout the experiments. Crack growth was automatically monitored using a video and image analysing system, coupled to a PC. Temperature during testing was 23°C ± 1 and relative humidity 50% ± 5. Specimens had been conditioned by exposing them to laboratory air of the same conditions for about 1 year. No precracking was done, so a crack initiation period might be present. However, as seen from Fig. 5.2, not much of this is seen in the crack length plot.

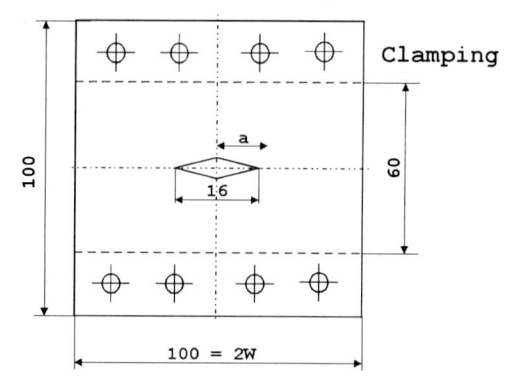

Figure 5.1   Crack propagation specimen, centre notched.

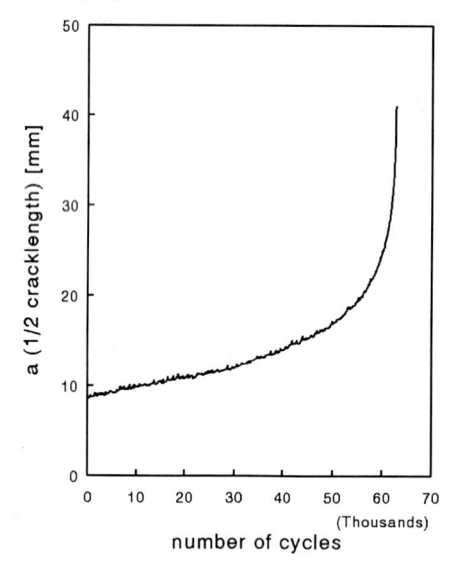

Figure 5.2   Example of the a-n curve as registered automatically using the image analysing system.

Because of scatter in the measured length of the crack, the crack growth speed was measured *by hand* from the crack length curves as shown in Figure 5.2. In this way 6 to 8 points remain for characterising the Paris curve. In Figure 5.3. the result is shown. It can be seen that the position of the Paris curve does not change when experiments are repeated.

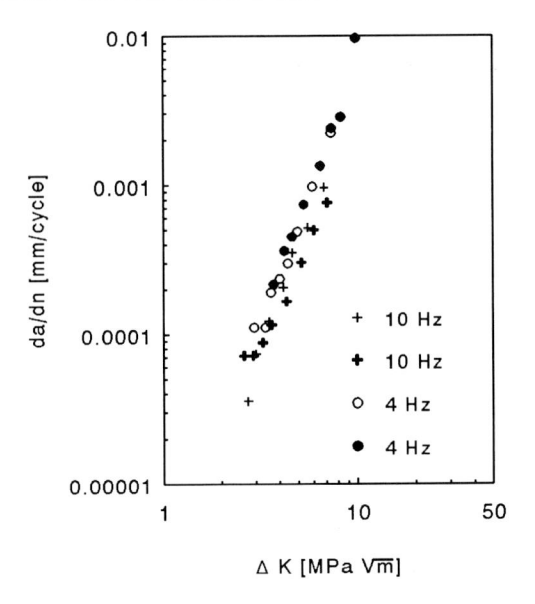

Figure 5.3   *Influence of frequency on FCP. 2 mm specimens. Loading direction transverse to MFD. Different symbols for each frequency indicate different experiments.*

5.3 Results and discussion

Four types of specimens were used, 2 and 5.75 mm thick plates, which were each tested with loading direction along MFD (Longitudinal specimens) or with loading direction perpendicular to MFD (Transverse specimens). Except for the case of 2 mm Transverse specimens, no difference in crack growth speed could be found between specimens tested at 10 and 4 Hz. For the case where a difference does exist, the crack growth speed for the specimens tested at 4 Hz are *higher* compared to the 10 Hz ones (Figure 5.4). This can be explained in two ways: The material used is time dependent, which is seen in higher Elastic Modulus and tensile strength at higher deformation rates. At 10 Hz the deformation rate is higher, so the material should perform better (§ 3.5.3). A second possible cause is heat build-up at the crack tip. At the higher frequency more energy will be dissipated per second, which will lead to a higher increase in temperature compared to the low frequency. If the temperature increase is confined to the crack tip area, *crack tip blunting* can be the result, leading to a decrease of stress intensity at the crack tip, and consequently lower crack growth speed [Lang,87a].

Because of the coinciding of results for different frequencies, in the overall results the average of experiments executed at both frequencies will be used.

Generally the Paris law is obeyed, as a linear relation exists between crack growth speed da/dn and stress intensity difference ΔK on log-log scale (see Figure 5.3). The relationship between crack growth speed da/dn and stress intensity difference ΔK can be written in the Paris law:

$$da/dn = A \cdot \Delta K^m \qquad 5.1$$

The stress intensity difference ΔK is equal to:

$$\Delta K = (\sigma_{max} - \sigma_{min}) \sqrt{\pi a} \cdot Y \qquad 5.2$$

a is half the crack length (See Fig. 5.1), σ the nominal stress, and Y is a geometry factor, which for this geometry can be approximated by:

$$Y = 1.015 + 0.0974(a/W) + 2.296(a/W)^2 + 1.010(a/W)^3 \qquad 5.3$$

W is half the width of the specimen, 50 mm.

A and m in the Paris law are the characteristic FCP parameters. However these are not material parameters, as the coefficients vary for specimens of different types. For each type of specimen results from at least 3 experiments were available. In the following the results of these are averaged. In Table 5.1 the Paris law parameters are summarised. In Figure 5.4 the crack growth rates for the four types of specimen are shown. As these are *averages* for three or more experiments, no individual points can be shown in these graphs. The position of the Paris Law curve is shown, and the interval in which this is valid (no measurements outside this interval of stress intensity difference ΔK are available).

Thickness	Orientation	A	m
2 mm	Longitudinal	$2.8 \cdot 10^{-7}$	3.55
2 mm	Transverse	$1.0 \cdot 10^{-6}$	3.83
5.75 mm	Longitudinal	$1.2 \cdot 10^{-6}$	3.05
5.75 mm	Transverse	$3.9 \cdot 10^{-7}$	3.87

Table 5.1   Parameters in the Paris Law, different specimen types

The 2 mm plate is much more anisotropic than the 5.75 mm plate, which shows here: the values for Transverse and Longitudinal loading are very different for the 2 mm plate, while these are very close for the thick plates. This reflects the results presented in § 3.4, the differences being caused by the relative sizes of core and shell layers. For the thick plates these are approximately equal, leading to similar properties in Transverse and Longitudinal directions. However, in the Longitudinal direction the variations in properties are higher over the width of the plate (Figure 3.14). So the material gets stronger with increasing crack length! In Figure 5.4 this is reflected in the smaller slope of crack growth speed against increasing ΔK, for the 5.75 mm thick Longitudinal specimens. In effect, due to this change in orientation the Paris curve need not be straight. The lines for all other experiments are approximately parallel.

As was mentioned in the introduction, methods exist to relate crack growth to mechanical properties. First the method of Wyzgoski [Wyzgoski,91 and 88] will be shown. Here the strain energy release rate ΔG is used. The relation between strain energy release rate and stress intensity is given in equations 5.4 and 5.5 for plane stress and plane strain situations respectively:

$$G = K^2/E \; ; \qquad \Delta G = \Delta K^2/E \quad for \quad R=0 \qquad 5.4$$

$$G = K^2/E \cdot (1-v^2); \qquad \Delta G = \Delta K^2/E \cdot (1-v^2) \quad for \quad R=0 \qquad 5.5$$

Chapter 5, Fatigue Crack Propagation experiments

These are valid for the *linear fracture mechanics theory* only. Wyzgoski used different values for E in both Longitudinal and Transverse directions. However as was seen in Chapter 3, the Modulus is not constant along the crack either. Especially in the thick Longitudinal specimen the stiffness increases when the side of the specimen is approached (Fig. 1.1). If a constant E along the crack length will be used, the resulting effect on the graph in Figure 5.4 will be limited to a decrease in angle by a factor of two for all specimens, and a different horizontal shift for all specimens. Therefore a variable Modulus along the crack length will be used, for those specimens where this variation was present. The results from the tensile tests presented in § 3.5.1 were used, and are summarised in table 5.2.

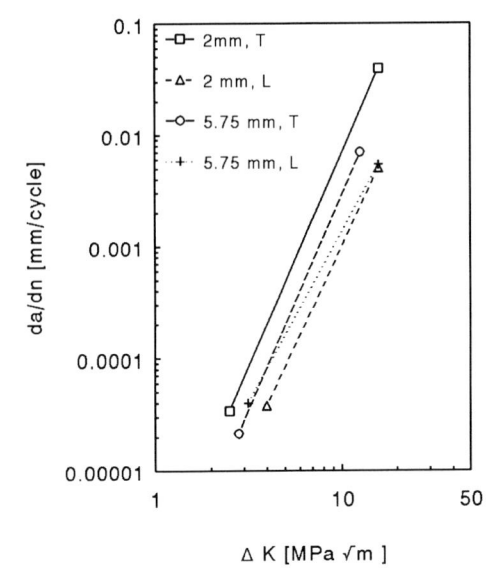

Figure 5.4 Crack growth rates for different specimen types. The lines represent the data points, while the length of the line indicates the interval over which the Paris Law applies (and measurements are available).

Thickness and orientation	Stiffness along crack length a	Min. and Max. Stiffness	Strength along crack length a	Min. and Max. Strength
2 mm, L	6815 + 24.4·a	7010 - 7495	137.8 + 0.36·a	141 - 148
2 mm, T	2820	2820	78.1	78.1
5.75 mm, L	4685 + 79.9·a	5400 - 7000	106.4 + 1.09·a	117 - 140
5.75 mm, T	5603	5603	114.1	114.1

Table 5.2 Summary of strength and stiffness values (unit: MPa). The min. and max. values are for the min. and max. crack lengths used to determine da/dn.

Using the Elastic Moduli shown above, and variable Moduli over the crack length when applicable, the calculation of ΔG converts Figure 5.4 into 5.5. Where before the lines for Transverse and Longitudinal 2 mm specimens were widely apart, representing the better alignment of fibres in the Longitudinal direction, now these have come practically together. Also the curves for the 5.75 mm thick specimens can hardly be distinguished from the 2 mm ones. In fact the differences are small compared with the natural scatter between individual experiments, compare with Fig. 5.3.

A modified Paris Law using this strain energy release rate can be written to represent the FCP behaviour [Wyzgoski,91 and 88]:

$$da/dn = C \cdot \Delta G^n \qquad 5.6$$

The constants are independent of geometry, and do not change with material thickness. The ΔG for plane stress (equation 5.4) was used in all cases, if the plane strain equation (5.5) would be substituted for the thick plates, assuming a Poisson's ratio ν of 0.35, G will be multiplied by a factor 0.88. In Figure 5.5 this will result in a shift of 0.056 to the left.

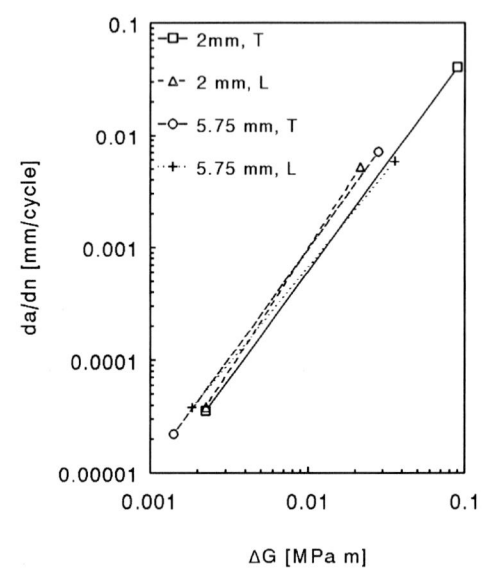

Figure 5.5   Crack growth rates against strain energy release rate, same experiments as in Figure 5.4.

A second method was used to relate the fatigue crack growth to the mechanical properties. This method is very much the same as the method used for lifetime measurements: the stress is normalised by the tensile strength, as in Chapter 4. The crack growth rate da/dn is plotted against the stress intensity difference, divided by the Ultimate Tensile Strength, Figure 5.6. That this procedure is actually equal to normalising the stress by the tensile strength can be seen if we rewrite equation 5.2, to yield equation 5.7, with in this case $\sigma_{min} = 0$:

$$\frac{\Delta K}{UTS} = \frac{\sigma_{max}}{UTS}\sqrt{\pi a} \cdot Y \qquad 5.7$$

Chapter 5, Fatigue Crack Propagation experiments 61

$$da/dn = A' \cdot \left(\frac{\Delta K}{UTS} \right)^{m'} \qquad 5.8$$

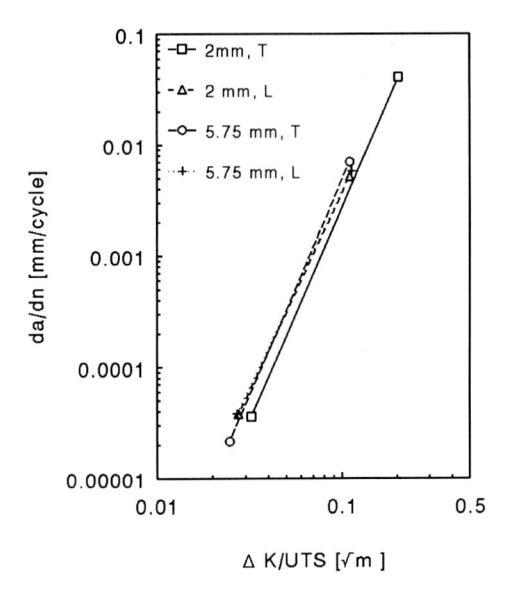

Figure 5.6   Crack growth rates against ΔK divided by UTS.

As can be seen in Figure 5.6 this second method is comparable to the strain energy release rate method in that the curves for the different orientations and different thicknesses are very close together after the normalising procedure. In equation 5.8 a modified Paris law is proposed, in which A' and m' are independent of fibre orientation within the specimen.

5.4 Conclusions

The method to normalise the stress intensity difference ΔK by tensile strength UTS satisfies well. Fatigue crack propagation behaviour can be characterised by one set of parameters, independent of fibre orientation and specimen thickness. Fibre orientation and mechanical properties vary over the plate used, and this variation must be included in the procedure to relate crack growth to mechanical properties, be it the Elastic Modulus or the tensile strength.

The fact that the Fatigue Crack Propagation rate can be related to local stiffness and especially local strength raises questions to the meaning of the results using the Linear Elastic Fracture Mechanics (LEFM) for calculation of the stress intensity at the crack tip. The inhomogeneity and anisotropy of the Elastic Modulus will *surely affect the stress distributions* over the width of the plate, and therefore the stress intensity at the crack tip. It can be concluded that simple LEFM, which yields for example the geometry parameter Y, is *not accurate* if the Modulus is not constant. Finite Element Methods (FEM) incorporating the variations of Elastic Modulus *should be used* to calculate the exact stress intensity at the crack tip. Apart from this, variations over the thickness of the specimens lead to a crack that is not straight through the thickness of the part, advancing first in certain layers.

The above inhomogeneity and anisotropy effect on the stress intensity at the crack tip is a different effect from that found by Lang et al. [Lang,84]. Here the effect was attributed to differences in the far field stress distributions for different types of specimen, resulting in differences in specimen stiffness and hysteretic heating of the specimens. The nonlinear visco-elasticity leads to Modulus variations across the specimen. However, in SFRTP's this effect will be secondary to the effect of the fibre orientation, because the glassfibres reduce strongly the visco-elastic behaviour of the material.

6. Microfoil Tensile Tests

6.1 Introduction

In this chapter an original method to measure fatigue damage is discussed. Micro Foil Tensile Tests (MFTT) are used to obtain strength profiles after a predetermined number of fatigue cycles. The decrease in strength and the change in fracture strain is measured, *in the individual layers* (core, shell skin). The result is an indication of the layers in the thickness where fatigue damage occurs first.

The MFTT method has been developed in our group at the Laboratory for Mechanical Reliability to determine degradation profiles of UV-degraded HDPE [DeBruijn,92].
The conventional technique for obtaining the strength or stiffness profiles of composites is by means of layer removal [O'Donnell,94]. These methods are not suitable for investigating strength profiles of fatigued specimens though. By milling from both sides, the strength of only one layer can be measured. The scatter in fatigue lifetime makes the construction of a strength profile of fatigued specimens impossible, as the percentage of fatigue lifetime for different specimens will vary. This is the main *advantage* of the MFTT method: All foils are cut from the same fatigued specimen, and thus have the same fatigue history. *Disadvantage* of the method is the damage that the microtoming process inflicts on the foils. Therefore an investigation was made of the variation of strength with foil thickness. Next the results of the strength profile measurements will be discussed. These profiles are compared with profiles for neat PA. Finally the strength profiles will be used to investigate the changes in properties due to the fatigue process. Changes in the profiles after fatigue indicate the location in the thickness where the damage exists.

6.2 Experimental procedure and materials

For the material and specimens used, see Chapter 4. From the fatigued specimens (the fatigue experiment was stopped before failure had occurred) and from not-fatigued reference specimens the thin foil specimens were obtained as depicted in Figure 6.1. For comparison also strength profiles were made for not reinforced Polyamide 6. This was supplied by DSM in the form of standard ISO R 527-1 test specimens of 4 mm thickness. A small dog-bone type specimen (ISO 6239-A2) was milled from the (fatigued) specimen, using a Roland PNC-3000 Computer Aided Modelling Machine. Specimen dimensions are given in Figure 6.2 From these miniature specimens thin foils of 85µm thickness were microtomed using a Leitz Microtome. To be able to hold the samples in milling and microtoming, the specimens are glued to a base. Sanding of the specimens and base are required for obtaining a reasonable bond. Both Araldite 2011 from CIBA Polymers and Kombi rapid from Bison were used. With both two component adhesives the problem occurred that the specimen sometimes debonded from the base during microtoming. To get the best possible bond the Araldite should be cured at 80°C for 30 minutes. As the curing process influences the material behaviour, see Fig. 6.9, the curing was omitted for the majority of specimens. Therefore the problem of debonding could not be solved. The maximum thickness was thus limited by the increasing microtoming force with increasing film thickness. Optimum foil thickness which was chosen was 85 µm.

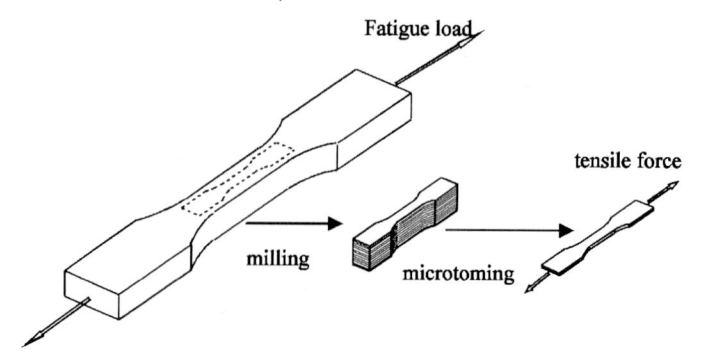

Figure 6.1   Steps in the production of MFTT specimens: From fatigued specimens small specimens are milled. Consequently from these the foils are cut by microtome.

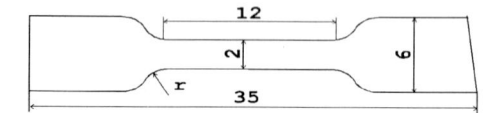

Figure 6.2   Dimensions of the MFTT specimen. r=3mm. The slant edge on the right is to facilitate microtoming (from right to left) Thickness: 85µm.

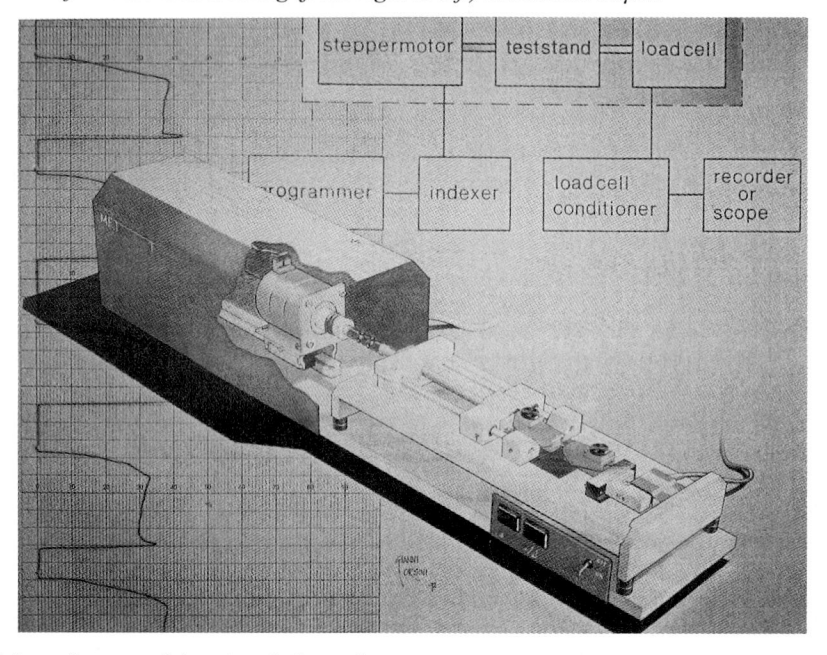

Figure 6.3   Lay out of the microfoil tensile test apparatus.

Chapter 6, Microfoil Tensile Tests for obtaining strength profiles

After microtoming the foils were stored in a wooden board containing slits in which the foils fit. This prevents curling of the foils. Directly after microtoming all the foils from a specimen (maximum 23 or 66 for the thin respectively the thick specimens), these were tensile tested in a specially designed test rig which was built in our laboratory, see Figure 6.3. The time between microtoming and testing was approximately equal for all foils. This prevents influence on the results from differences in water absorption between foils, due to different exposure times to air.

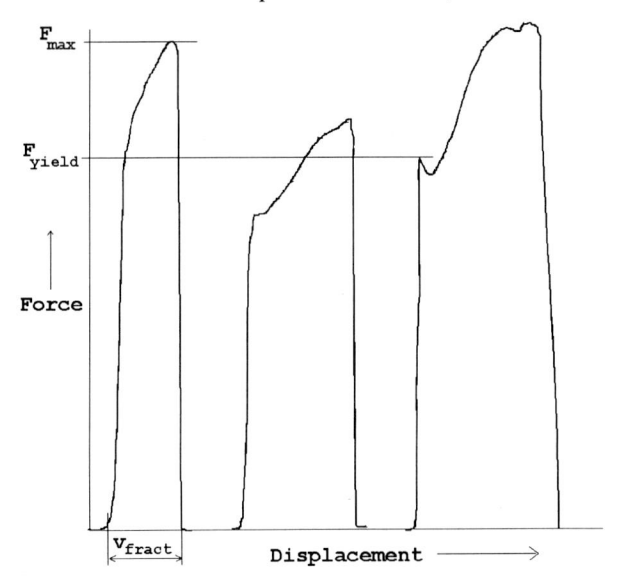

Figure 6.4   Force - Displacement curves for 3 cases: Left: GFPA, longitudinal direction, Shell. Middle: GFPA, Core. Right: Unreinforced PA6.

The length between clamps was 17 mm, and the clamp separation speed was 0.3 mm/s. This led to a strain rate comparable to that in tensile testing of the normal specimens depicted in Fig. 3.12. In Fig. 6.4 examples are given for the tensile curves generated by MFTT. Differences between reinforced and non reinforced Polyamide are mainly seen in the occurrence of a yield point in the latter case. For the foil from the core layer, containing mainly transverse fibres (Longitudinal specimen), the form is similar to the non reinforced Polyamide, though the yield is much less pronounced. For foils from the shell layer, containing fibres mainly aligned with loading direction, no yield point exists. The fracture strain for the fibre reinforced PolyAmide is smaller compared to the non reinforced grade, and smaller for aligned than for perpendicular fibres.

6.3 Microfoil Tensile Test profiles

The first series of measurements were for the purpose of establishing the method only. Various approaches were followed: The first was to measure the variation of foil strength with thickness of the foil. The second was the evaluation of foil thickness after microtoming. The foil thickness gives some information about the fibre orientation. Last method to evaluate the reproducibility of the method was the comparison of strength profiles from basically equal specimens.

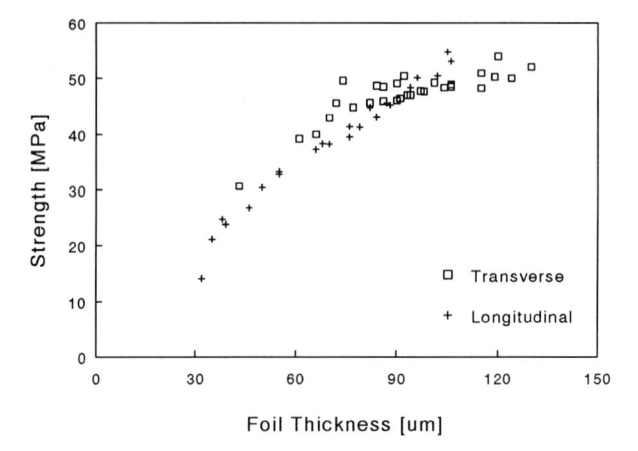

Figure 6.5   Variation of foil strength with foil thickness.

Variation of foil strength with foil thickness
For various types of specimens the variation of strength with foil thickness was evaluated. In Figure 6.5 a typical result is shown. For foils from both transverse and longitudinal specimens the strength increases with foil thickness, until a thickness of around 100 μm. For foils of higher thickness the strength is almost constant.

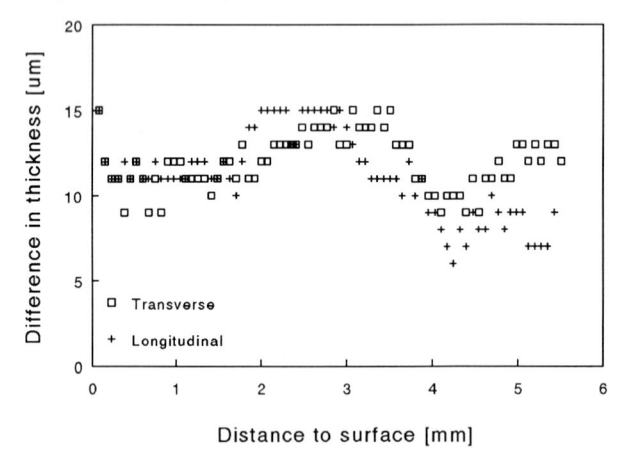

Figure 6.6   Difference of measured foil thickness with thickness setting of microtome. The effect of the Core layer is seen.

Variation of foil thickness over the specimen thickness
When thickness was measured after microtoming, it appeared that the foils were thicker than the cut thickness. The difference was in the order of 5 - 20 μm. The difference was not constant, as can be seen in Figure 6.6; in the core layer the difference is larger compared to the shell layers. SEM photographs of the cut surface of the foil showed the reason for the difference between cut thickness and measured thickness (Figure 6.7). The matrix is smeared over the surface. Fibres

Chapter 6, Microfoil Tensile Tests for obtaining strength profiles   67

are pushed elastically into the foil during microtoming and, after the passage of the microtome knife, come out again, see Figure 6.8.
The measured thickness obviously does not represent a valid measure for the thickness that is actually loaded. Therefore the cut thickness was used to calculate the stress in the foil specimens. The fact that the difference between cut thickness and measured thickness is not constant, is caused by the misalignment of the fibres from the planar orientation, see paragraph 2.2. This misalignment is low at the surface and in the shell layer, and increases towards the centre of the specimen, especially for the thick, 5.75mm, specimens. As the microtoming takes place parallel to the surface, these fibres must be cut during microtoming. The cutting of these non-planar fibres leads to an increase in difference between cut thickness and measured thickness.

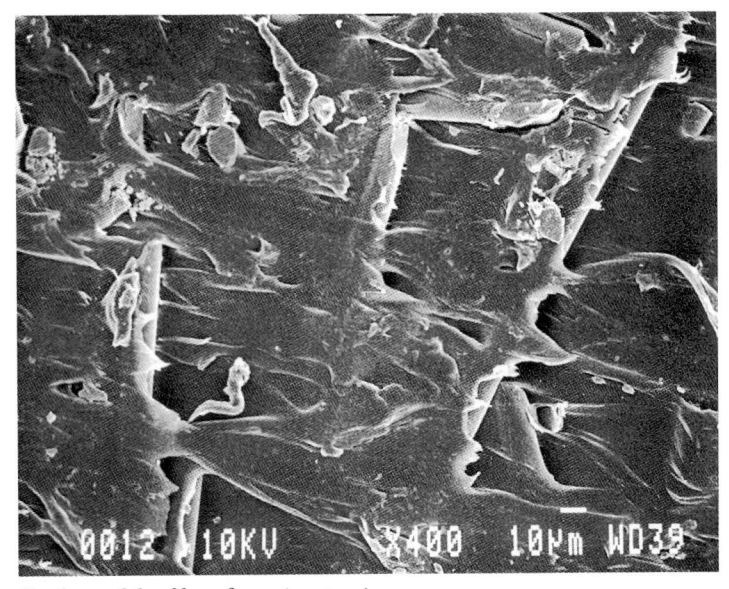

Figure 6.7   Surface of the film after microtoming.

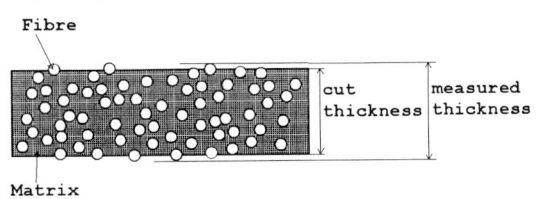

Figure 6.8   Explanation for the difference between cut thickness and measured thickness of a microtomed foil.

Scatter in strength profiles
To evaluate the scatter from profile to profile, strength profiles of 3 not fatigued specimens were measured. As the fatigued specimens are compared to these not fatigued reference specimens, knowledge of the possible scatter is of great importance. In Fig. 6.9 the strength and fracture strain profiles for three CL-type MFTT specimens are shown. References 1 and 2 are identical while reference 3 was cured for 30 minutes at 80°C to cure the Araldite adhesive.

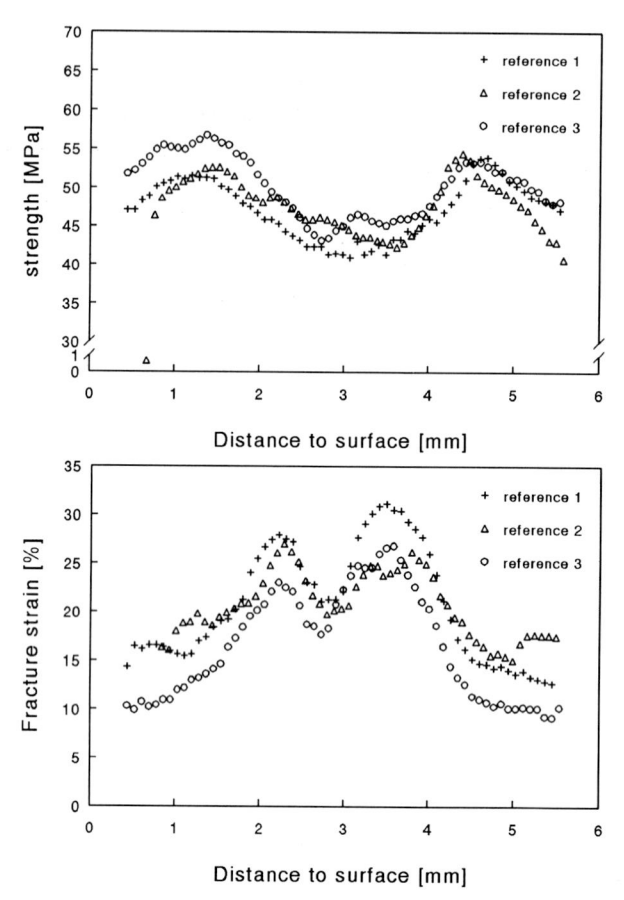

Figure 6.9 Strength and fracture strain profiles for 3 not-fatigued CL-type specimens. Reference 3 has been cured for 30min. at 80°C.

The maximum difference amounts to 4 Mpa (between reference 1 and 2), this is 8 - 10% of the strength. As is visible in Fig. 6.9 the thicknesses of shell and core layers are the same for all specimens, the strength in these layers is not. References 1 and 2 reproduce quite well, reference 3 differs significantly from these, which can be accounted for by the curing at 80°C. This will have decreased the water content in this particular specimen. In agreement with the higher strength of reference 3 is the lower fracture strain for this specimen. The maximum difference in fracture strain is 4%, between reference 1 and 2. Unknown is if the difference in strength profile between references 1 and 2 is an artifact from the MFTT procedure, or if it is a real difference in material behaviour. The MFTT scatter can be compared to the scatter that is found normally in the tensile strength of full size specimens. The latter is quite small, 2 - 3% is found in almost all series of tensile experiments. However this does not prove that the scatter found in the MFTT results is an artifact, the possibility exists that local variations in fibre orientation or fibre content cause this.

6.3.1 Form of the MFTT profile, influence of matrix

The form of the strength profile, as measured using the MFTT technique, can be characterised for almost all specimens by 3 zones, see Fig. 6.10. Generally at the middle of the specimen the core layer has a low strength, further from the middle the strength increases gradually, to a maximum value in the shell layer. Towards the surface of the specimen the strength decreases, but generally not to as low a value as measured for the core. The thickness of the core layer characterizes the type (location in the plate) of the specimen measured. The form of the profile is *not caused by the fibre orientation alone*, what might be expected. This can be seen in Fig. 6.11 where the strength profile for not reinforced PA6 is shown.

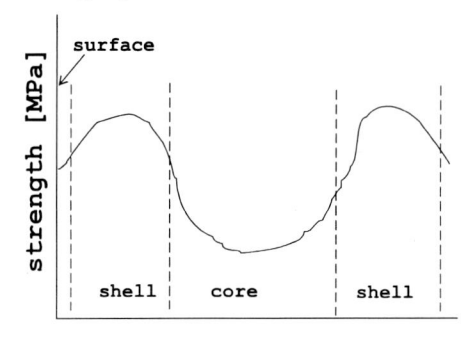

Figure 6.10 General form of strength profiles of GFPA, when measured in the longitudinal direction (MFD).

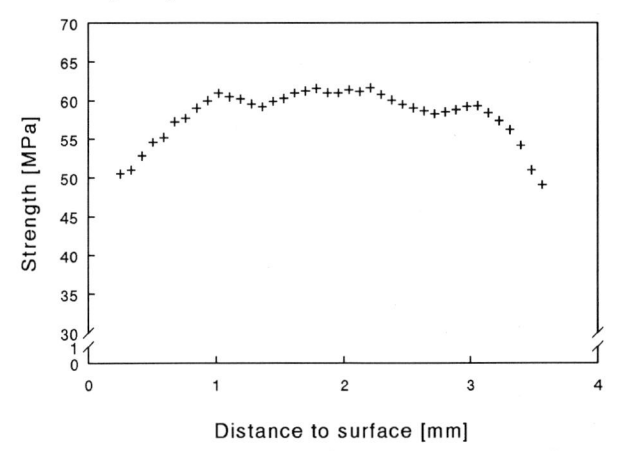

Figure 6.11 Strength profile for unreinforced PA6.

The profile for the unreinforced PA6 of Fig. 6.11 shows an approximately constant strength in the middle, and a gradual decrease of strength towards the surface, over approximately 1 mm. When viewing the strength profiles of reinforced PA6, we observe a similar decrease towards the surface, over 0.5 - 1 mm. That this decrease in GFPA can not be explained by the skin layer, with less aligned fibres compared with the shell layer, can also be deduced from the thickness of the skin layer, which is only a few tenths of a millimeter. O'Donnell and White [O'Donnell,94] did measurements in elastic bending, using the layer removal method to obtain a profile of Elastic Modulus. For both unreinforced and glassfibre reinforced PA6.6 a decrease in Elastic Modulus towards the surface of the specimen was found. In this case the decrease occurred in a layer of approximately 0.5 mm. Measurements of fibre orientation did not show any significant change in this area. O'Donnell's conclusion therefore was that the decrease in stiffness towards the surface in the reinforced PA6.6 is caused by a decrease in matrix stiffness. O'Donnell does not give an explanation for this decrease of stiffness towards the surface, three characteristics of the polymer can change however, and explain the change both in stiffness and in strength: 1) Molecular orientation [Folkes,80] of the polymer, 2) degree of crystallinity [Paterson,92], 3) Differences in water content. The molecular orientation (if of importance) would increase the strength and stiffness of the polymer towards the surface, where the higher molecular orientation (in Mould Flow Direction) is present. Apart from the *degree of crystallinity* being lower at the surface of the moulding, also in conditioned Polyamide the *water content* at the surface can be higher than in the centre of the moulding. Paterson and White [Paterson,92] found a decrease in T_g towards the surface for conditioned material, while for dry as Moulded material this decrease was absent.

For the not reinforced material used, a change in morphology could be observed using Polarised light on a normal microscope. However this change was at a depth of 0.3mm from the surface. The decrease in strength towards the surface for this material is over a depth of 1 mm, see Fig. 6.11. Therefore the conclusion is that *variations in water content* must be the main cause for the decrease in strength towards the surface of the specimens.

6.3.2 Influence of fibre orientation on MFTT profiles

The strength profiles of course depend strongly on the fibre orientation in the specimens. Generally the same trends are expected as in *full-thickness specimens*: Orientation of the fibres in the testing direction will give a high strength and a low fracture strain, while perpendicular orientation will give a lower strength and a higher fracture strain. A distinction must be made between thin (2mm) and thick (5.75mm) specimens, as the strength profiles are essentially different.

6.3.2.1 Profiles of thin (2mm) specimens

Figure 6.12 shows strength and fracture strain profiles for specimens taken from the middle of the plate (see Fig. 6.13), both in <u>Longitudinal and in Transverse direction</u> (respectively parallel and perpendicular to MFD). A small effect of the core layer is seen, where the fibre orientation perpendicular to MFD gives a *low strength* when the specimen is in Longitudinal direction, and an *increase* in strength in the core layer is seen in the profile for the specimen in transverse direction. The strength and fracture strain of the core layer measured in transverse direction is not equal to the values for the shell layer in longitudinal direction: The strength in the shell layer is lower, and the fracture strain is higher, indicating less alignment of the glassfibres. The effect of the higher water content near the surface is visible in the lower strength towards the surface.

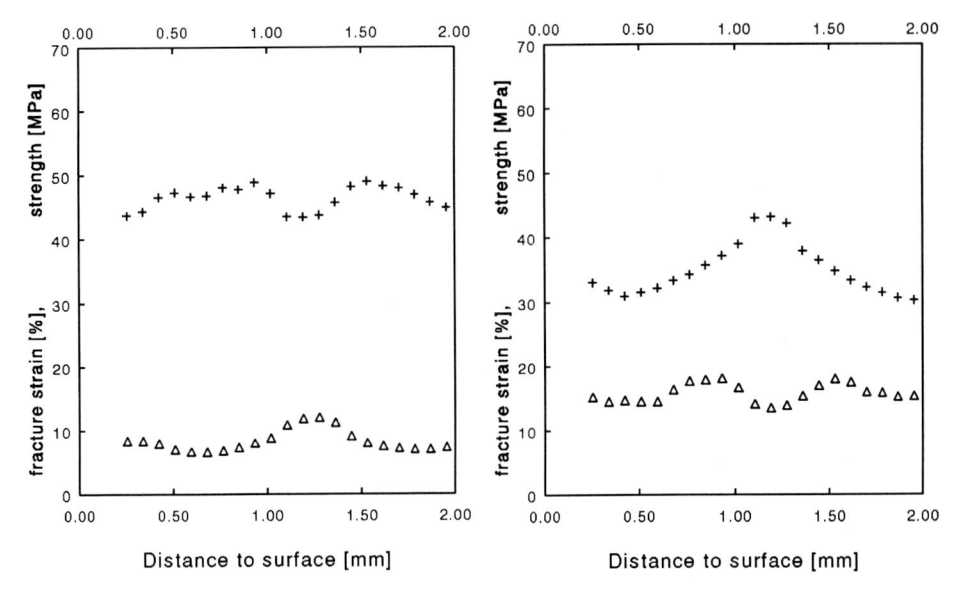

Chapter 6, Microfoil Tensile Tests for obtaining strength profiles

Figure 6.12   Longitudinal (left) and Transverse (right), CL respectively CT specimen profiles.

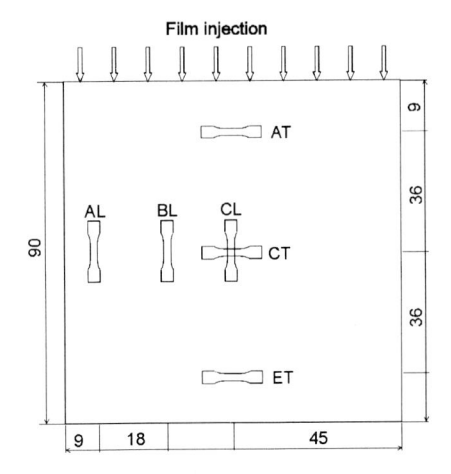

Figure 6.13   Location of the MFTT specimens in the 90 x 90 mm plate of 2mm thickness.

<u>Over the width of the plate</u>, profiles of AL, BL and CL type specimens (see Fig. 6.13) have been measured. The plate is assumed to be symmetrical, with DL and EL specimens being similar to BL and AL ones respectively. Results are shown in Fig. 6.14a. While the CL profile shows the lower strength in the core layer, associated with the orientation of the fibres being transverse to the loading direction, the strength of the BL type shows much less effect of the core. In the AL type, located at the side of the plate, this effect is almost absent. In the fracture strain, the effect of the core layer is more clearly visible, although less so for the AL type specimen.

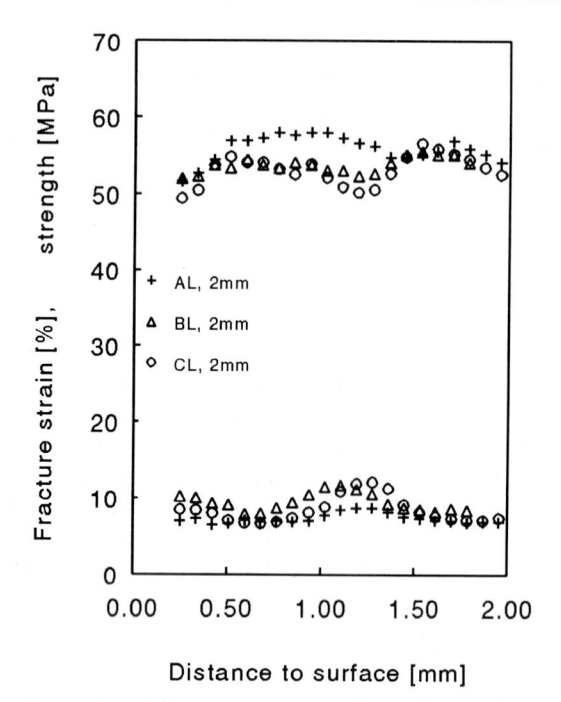

Figure 6.14a Strength and fracture strain profiles in longitudinal direction. 2mm.

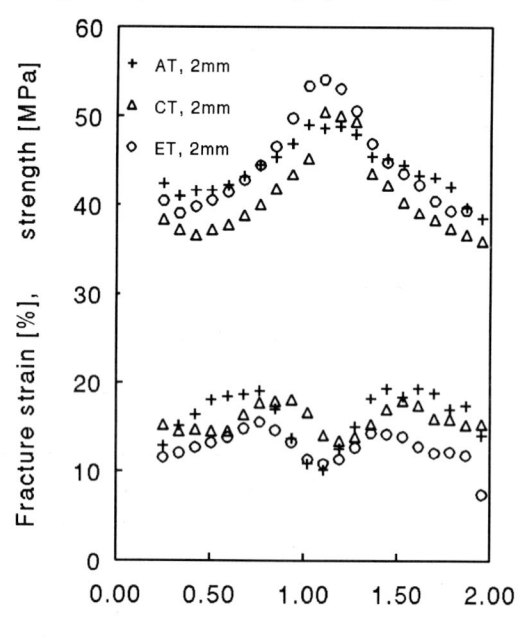

Figure 6.14b Strength and fracture strain profiles in transverse direction. 2mm.

Profiles for the transverse specimens, Fig. 6.14b, are all very similar. In all specimens the effect of the core layer, with higher orientation in the testing direction is visible, both in the strength as well as in the fracture strain. The effect of the distance to the gate is small. Close to the gate and far from the gate the strength is a little higher compared to the centre of the plate. The fibre orientation in Longitudinal direction increases with increasing flow path, which corresponds with what was found by Bright et al. [Bright,78]. This explains the lower strength of the CT type compared with the AT type specimen. Close to the end of the moulding the flow changes, resulting in a more transverse orientation of the fibres, and a higher strength of the ET specimen.

6.3.2.2 Profiles of thick (5.75mm) specimens

The strength profiles for the thick specimens show some differences with the profiles for the thin plates, reason why these have to be discussed here separately. The behaviour we expect to see is that of the thin specimens: the longitudinal specimens should show strong shells, and a weaker core layer, while transverse specimens should show a core layer that is stronger than the shell layers. In Fig. 6.17 it can be seen that this is not the case! In Fig. 6.16 the fibre orientation at the DT location (see Fig. 6.15) is shown. In the longitudinal direction (DTL profile) fibres in the Shell layer are oriented parallel, while in the Core layer the fibres are perpendicular. In the transverse direction this inverts, of course: In the Shell fibres are perpendicular, while in the Core these are parallel to the loading direction.

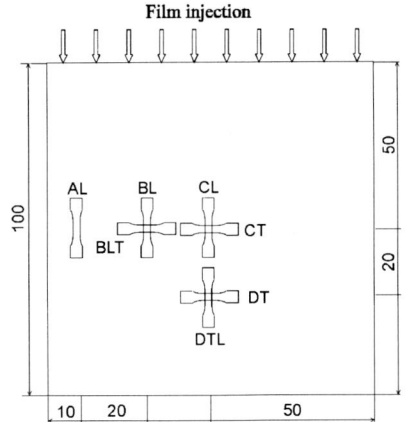

Figure 6.15   Locations of the MFTT specimens in the 100 x 100 mm plate, 5.75mm thickness.

At the DT location, the MFTT profiles show strength profiles which are very similar in both longitudinal and transverse directions, showing a lower strength in the core layer, compared to the shell layers. The core and the shell layers do behave differently though in both directions, which can be seen in the fracture *strain*, Fig. 6.17. Here the expectations are met: *high fracture strain in the case of perpendicular fibres*, and a *low fracture strain for fibres parallel* to the foil. The orientation distribution can explain the discrepancy between the expected and measured results, in the case of the transverse measurements. The higher strength, that we would have expected to occur in the core region, is not present, because the fibre orientation in the core is not a planar orientation: fibres exist especially in this layer, that are are not parallel to the surface of the moulding: fibres with orientation at an angle to the surface of the specimen are present in the core area, and weaken the foil.

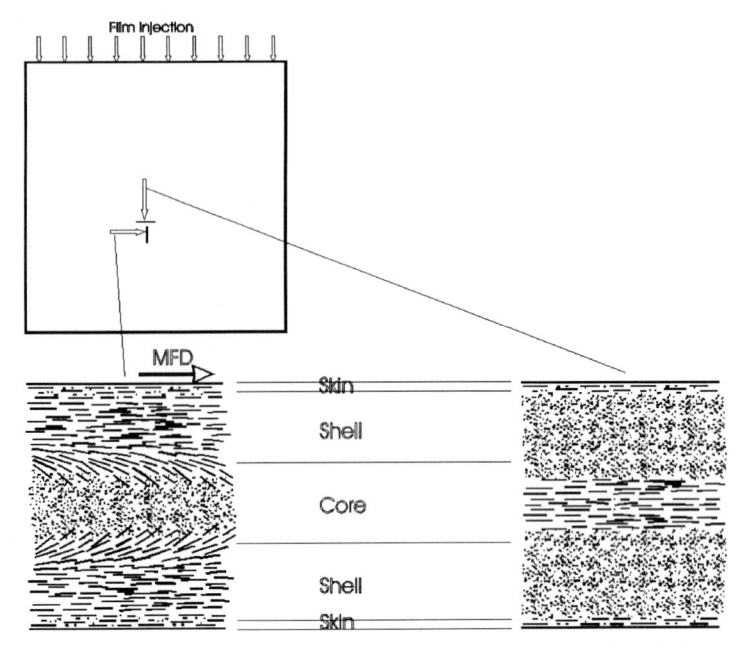

Figure 6.16 Fibre orientation at DT location, seen in the longitudinal and the transverse direction. Fibres with a non planar orientation are not recognised on a transverse cross-section (right).

In Fig. 6.17 the orientation and the measurements of the foils are joined, showing the relation between the two. In the orientation graphs at the top of Fig. 6.17, the cutting direction of the foils is vertical. The foils can now be divided in 5 classes, according to their fibre orientation. Their strength and ductility behaviour can be explained.

1   Low fracture strain (<10%) occurs when the majority of the fibres are oriented in the direction of the load.

      a: If all fibres are well aligned in the loading direction, the UTS is relatively high (55-65 MPa) (Shell layer, Longitudinal foil)

      b: If some fibres are not well aligned, the foil is weakened (ca. 55-60 MPa), not reaching the same high UTS as with aligned fibres. The low fracture strain remains. (Core layer, Transverse foil)

2   Intermediate fracture strain (ca.15%) occurs when the fibres are oriented perpendicular to the load.

      a: This fibre orientation leads to a relatively high strength of the foils (55-60 MPa), when all fibres are aligned. (Shell layer, Transverse foil)

Chapter 6, Microfoil Tensile Tests for obtaining strength profiles 75

b: The strength of the foils is drastically reduced (ca. 50 MPa) when some of the fibres stick out of the microtomed foil. (Middle of the Core layer, Longitudinal foil)

3 High fracture strain occurs (>20%) when the fibres are oriented at an angle to the axis of the foil, leading to a relatively low strength (50-60 MPa). Probably a shear deformation mechanism occurs, which leads to the high fracture strain. The fibres have a tilt angle to the microtomed plane and are cut during microtoming, inducing damage and consequently low strength. (Core layer, Longitudinal foil)

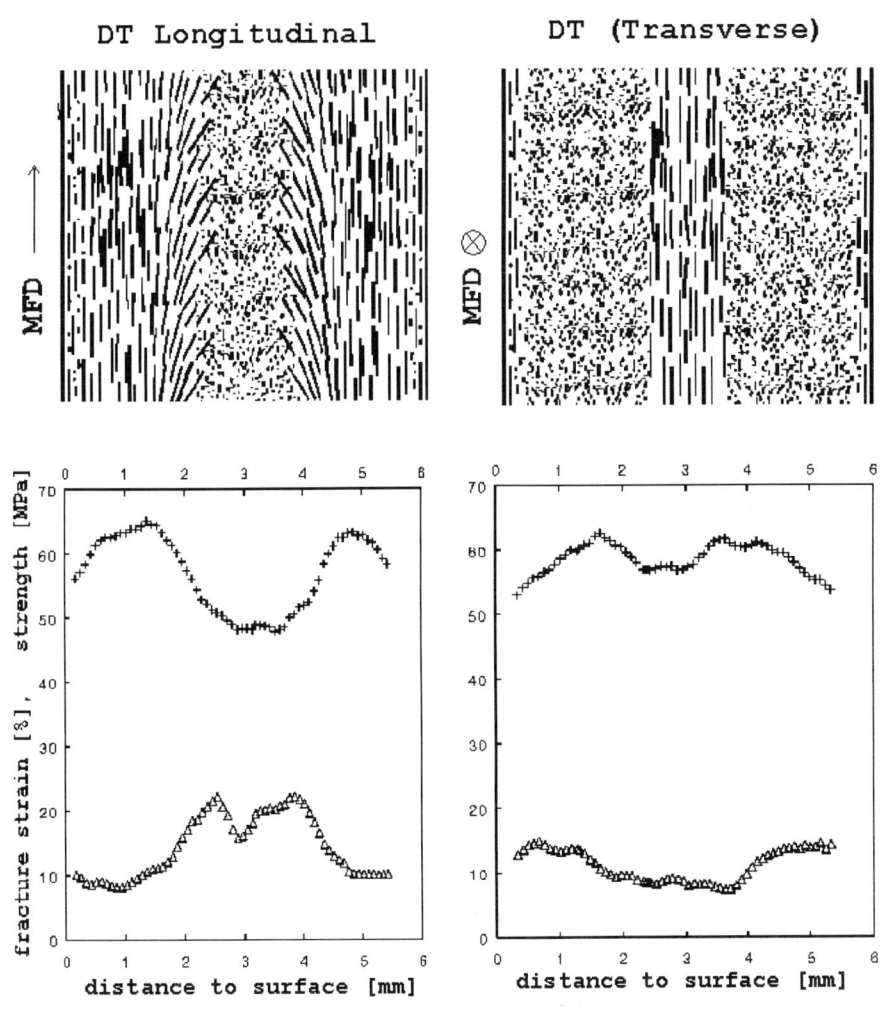

Figure 6.17 Comparison of strength and fracture strain profiles in both longitudinal and transverse (to MFD) directions. The tensile tests on the foils were executed in the vertical direction of the orientation graphs of the cross section depicted at the top.

The gradual decline of the strength towards the surface is, as was mentioned before, not caused by the fibre orientation, but by the decrease in matrix properties due to water absorption, as was proven by using unreinforced PA6 (§ 6.3.1).

One striking feature, visible in the strength profiles of these thick specimens is the asymmetry, a feature that will be hard to show using any other method. The asymmetry of the profiles indicates an asymmetric flow during filling. This can be caused either by the asymmetric gate, which is placed not exactly at the centre of the thickness of the plate, or more likely by asymmetric cooling. If the temperature at the mould walls during filling is not the same, one frozen layer (see Fig. 2.2) will grow faster than the other. The temperature profile over the melt will not be symmetric, and thus the velocity profile will not be symmetric.

Variation of profiles with position in the plate
Seven types of specimens were investigated. These were cut from the original square plates at different locations and in different directions, as indicated in Fig. 6.15. Type names are identical as in the case of the full-size specimens, except when the location of the specimen and the orientation do not match: A specimen at the location of the DT specimen, but tested in the Longitudinal direction is called a DTL specimen. As the transverse specimens show similar strength patterns as the longitudinal ones, only the last are shown here. In Fig. 6.18 AL, BL and CL profiles are shown. Obvious is the decrease in core thickness, when moving from the centre of the plate (CL type) to the side. This decrease in core layer thickness is the main cause for the differences in strength of the full-thickness specimens, as found in paragraph 3.4.

Also the strength in the shell layers is higher in the case of the AL specimen, indicating a better alignment of the fibres. When comparing to the profiles for the thin specimens (Fig. 6.14a) it is obvious that the differences between shell and core layers are much more pronounced for these thick specimens.

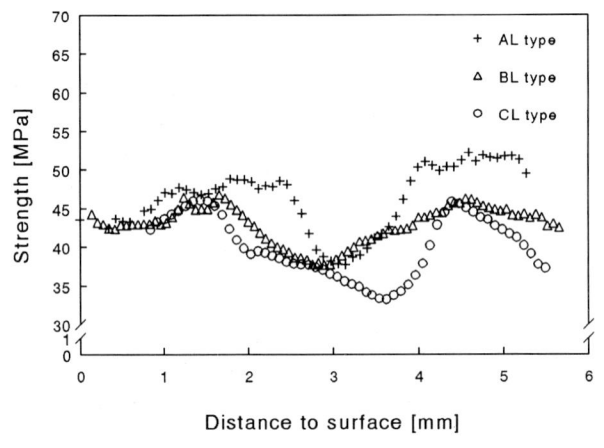

Figure 6.18   Variations of strength profiles with location in the plate.

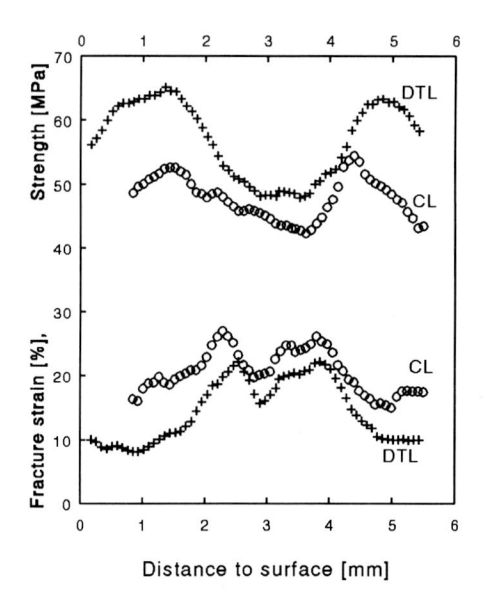

Figure 6.19   CL respectively DTL profiles. The CL profile is closer to the gate.

Figure 6.19 shows the influence of *distance to the gate*. As can be seen in Fig. 6.15 the CL specimen is closer to the gate. The alignment of the fibres increases with flow path, resulting in higher strength and lower fracture strain for the DTL specimen. This corresponds with what was found by Bright et al. [Bright,78]. The asymmetry of the profiles for both specimens is very similar, over the flow path little changes take place.

6.4 Influence of fatigue on MFTT profiles

In a preliminary research the MFTT method showed some promising results in detecting fatigue damage in short glassfibre reinforced PA6 [Horst,96a and 95a]. As was shown in the previous paragraphs of this chapter, evaluation of the strength and fracture strain profiles which result from the method is not as straightforward as one would wish. In § 6.3 it is shown that possible scatter in strength is 4MPa, and in fracture strain 4%. Therefore the differences between a fatigued specimen and the unfatigued reference must be higher than these values, to be able to conclude that damage is present.

Quite some profiles of fatigued specimens have been made, both for 2 mm as well as 5.75 mm thick specimens. In the experiments 3 factors have been varied:
1   Fibre orientation distribution through using different specimen types.
2   Fatigue load level.
3   Fatigued life (percentage of expected fatigue lifetime).
Other factors were kept approximately constant, like the water content. All specimens used were conditioned to equilibrium in air.
The three dimensional matrix of possible experiments can not be explored entirely of course. In

the preliminary experiments for a few specimen types some profiles were made for fatigued thick specimens. For CL type, thick specimens both the fatigue load level and the percentage of fatigue lifetime was varied. For thin (2 mm) specimens damage was measured for various specimen types, though all were fatigued at the same load level, and at a high percentage of predicted lifetime, generally as close to failure as possible.

6.4.1 Effect of number of cycles and load level

For investigating the development of fatigue damage during fatigue, the damage was measured at various percentages of predicted lifetime. To be able to predict the fatigue lifetime accurately, at each load level *10 fatigue experiments* were executed, to obtain a reliable relation of creep speed with number of cycles to failure This according to the creep speed method presented in 4.3.3. At each load level the influence of percentage of lifetime was measured, to investigate if fatigue damage develops differently at different load levels. One specific specimen type was used for this investigation, the CL specimen, a longitudinal specimen from the middle of the plate. Fatigueing was done following the standard method used in this investigation, see Chapter 4. Experiments were done at 55, 60, 70 and 80% of UTS, average lifetimes are shown in Table 6.1.

Load	Average lifetime	Min. Lifetime	Max. Lifetime	Error using creep speed	Error using S-N curve
0.8UTS	230	170	300	20	60
0.7UTS	630	460	870	45	200
0.6UTS	5400	4100	6800	800	1400
0.55UTS	25,000	15,000	36,000	2200	10,600

Table 6.1   *Lifetime results for CL type fatigue experiments. Maximum possible error is indicated, as based on the outer bounds of the scatterband, both when the creep speed method is used, and if the lifetime would be estimated from the S-N curve alone.*

Using the creep speed method decreases the uncertainty in the predicted lifetime for the cases where the fatigue experiments are stopped in order to measure the strength profiles. The remaining error is approximately 10% maximum. In Fig. 6.20 an example is shown for the fatigue lifetime - creep speed graphs.

General results: CL type specimens
At all load levels a decrease in strength over the entire thickness is seen, if the specimens have been fatigued sufficiently, Fig. 6.21. This is accompanied by a decrease in fracture strain, especially in the layer where the fracture strain was high before fatigueing, the core. When in the strength profile no significant change can be seen, it is possible that a decrease in fracture strain occurs. This is visible in Fig. 6.21 for the specimen fatigued at 0.8UTS. Both in the shell as well as in the core layer a decrease in fracture strain, compared to the reference specimen is visible.

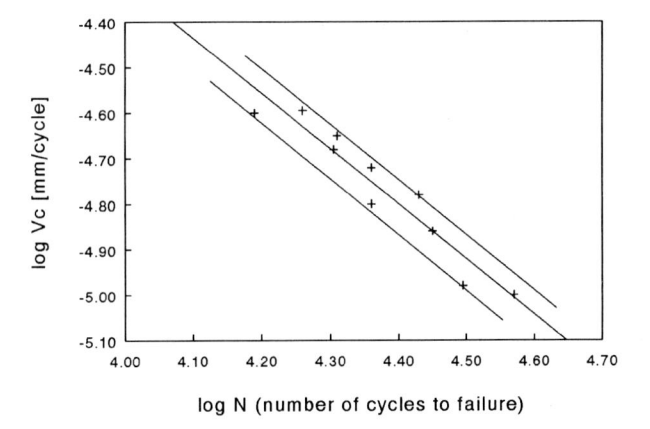

Chapter 6, Microfoil Tensile Tests for obtaining strength profiles

Figure 6.20 Graph of creep speed with number of cycles to failure. CL type specimens fatigued at 55% of UTS. The middle line represents the regression line, the outer lines indicate the outer bounds of the scatterband.

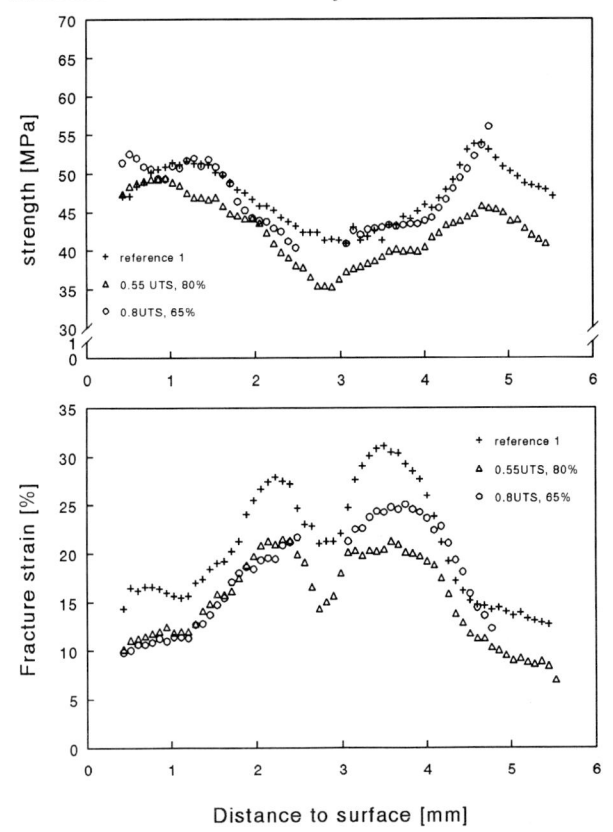

Figure 6.21 Examples of fatigued CL type specimens. One fatigued for 80% of predicted lifetime, at 55% of UTS, the other for 65% of predicted lifetime, at 80% of UTS.

Only at low fatigue loads (0.6UTS or lower) a specific *local damage* is noted. At the location of stress whitening lines, the foils show *low strength* and *low fracture strain*, see Figure 6.22. Specimens fatigued at higher loads do show the stress whitening lines, although no change in behaviour of the foils is noted. In Figure 6.22 it can be seen that the local decrease in strength can occur either in the shell or the core layer, though the phenomenon was seen more often in the core layer. The decrease in strength is accompanied by an extreme decrease in fracture strain, to a level of 10 % or lower.

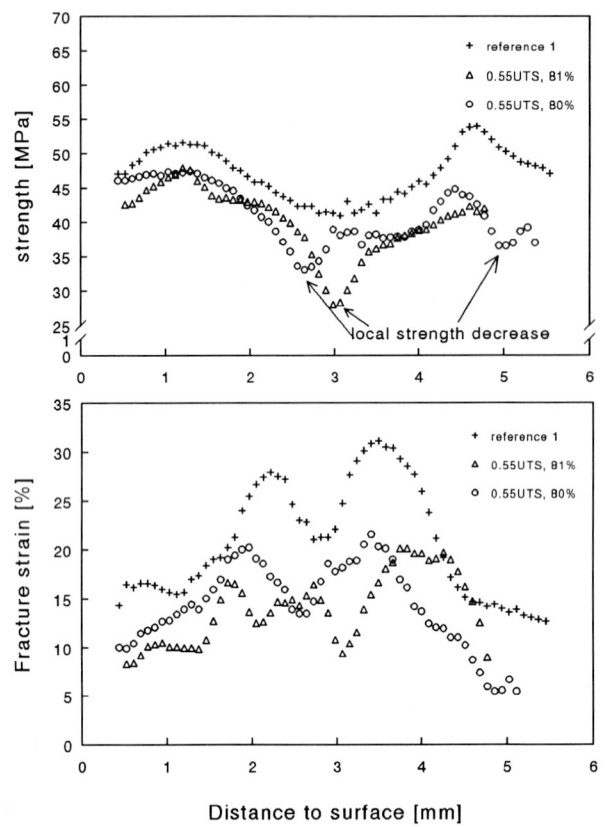

Figure 6.22   Examples of local damage in the CL type specimens, after being fatigued at 0.55% of UTS. For both the experiment was stopped when 80% of predicted lifetime was reached.

Influence percentage of lifetime, CL type specimens
For specimens fatigued for less than 80% of the predicted lifetime, no significant change in the MFTT strength profiles of the fatigued specimens, compared with the reference can be noted. For specimens fatigued up to a higher percentage of predicted lifetime, generally a decrease of strength can be seen. However, it is not possible to quantify the increase of damage with increasing percentage of fatigue, as the predicted lifetime shows an error of 10%.

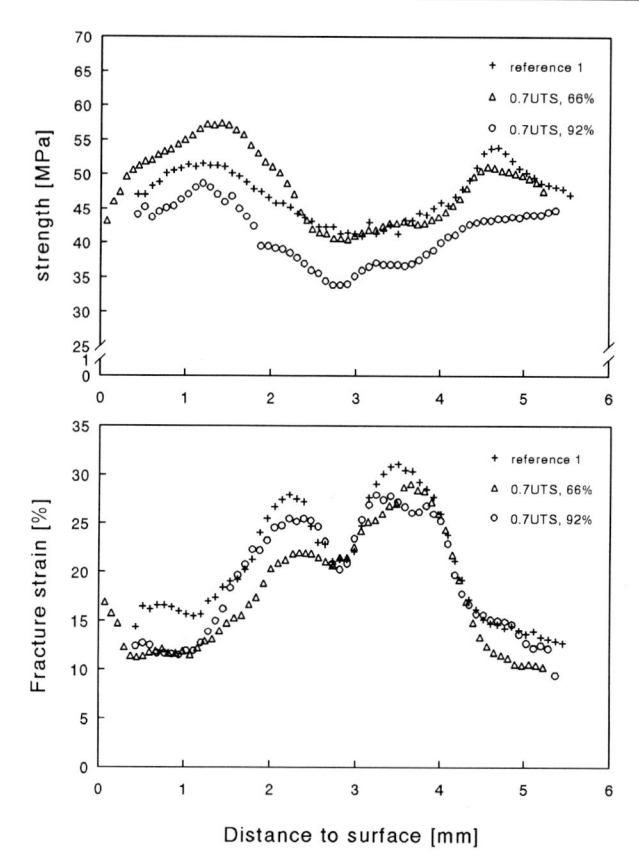

Chapter 6, Microfoil Tensile Tests for obtaining strength profiles 81

Figure 6.23 Examples of influence of percentage of fatigue life on damage in the CL type specimens, after being fatigued at 0.7% of UTS, for 66 respectively 92% of predicted lifetime.

In Figure 6.23 the development of damage with fatigue is illustrated. The decrease in strength for the specimen fatigued up to 92% of predicted lifetime is obvious, while for the specimen fatigued up to only 66%, no decrease in strength is seen. In one shell layer the fatigued specimen is stronger, compared to the reference. The higher strength is accompanied by a low fracture strain over the same part of the thickness. No explication can be given for this phenomenon.

Thus above 80% of predicted lifetime the fatigue damage affects the strength profiles. At lower percentages the fracture strain of the specimens can be affected. The scatter in experimental results, especially from one profile to another (not within one profile) makes it impossible to quantify the amount of damage.

Influence of fatigue load level, CL type specimens
At all load levels a decrease of strength and a decrease in fracture strain in the MFTT profiles can be observed, provided that the specimens have been fatigued long enough, for more than 80% of their predicted lifetime.

The local drops in strength, at the location of stress whitening lines, were observed *only for* the specimens fatigued at *loads lower than or equal to 60% of UTS*.

The amount of decrease in strength that can be observed in the strength profiles for the fatigued specimens, shows little correlation with the fatigue load level. The fracture strain **does** show a more clear correlation. The fracture strain **decreases** more with decreasing fatigue load level, and increasing fatigue lifetime. In Fig. 6.24 results for all experiments are summarised. The upper values represent the strength in the stronger (shell) layer, and for the fracture strain in the core layer, the lower values being the values for the core and shell layers respectively. The lines between values indicate the scatter in results for the different specimens, fatigued at one load level, but fatigued up to different percentages of predicted lifetime.

The damage in these CL type specimens does show more clearly in the fracture strain profiles, than in the strength profiles. A decrease of strength is seen only in the (stronger) shell layer, while in the core layer only occasionally a decrease in strength is found, mainly at the location of stress whitening lines. The fracture strain does show a clear decrease, with decreasing fatigue load level.

Final conclusion is that the *damage increases* if the specimen is fatigued for a *longer number of cycles*, at a *lower load level*.

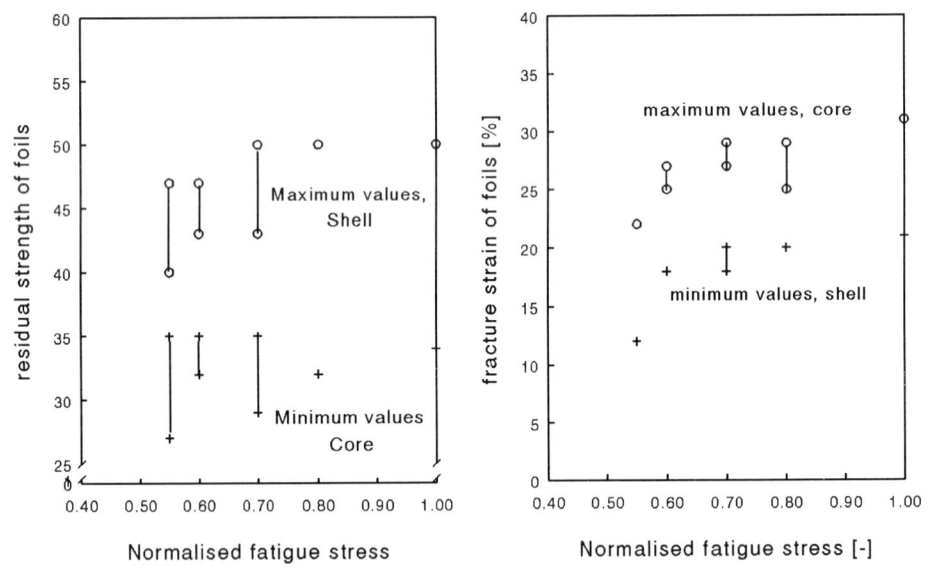

Figure 6.24 Relationship between fatigue load level and decrease in properties. At normalised fatigue stress = 1 the values for the not fatigued reference are plotted. Note that the fracture strain *is more sensitive to fatigue compared to the residual strength*.

6.4.2 Effect of specimen type

Now that the effect of fatigue load level on the one hand, and percentage of fatigued life on the other was known, a series of measurements on different specimens was performed. Goal of this investigations was to get some indication about the layer in which fatigue damage is most likely

Chapter 6, Microfoil Tensile Tests for obtaining strength profiles 83

to occur. Of course this may be also in both core and shell layers. As the effect of the fibre orientation is different for 2 mm and 5.75 mm specimens, which was shown in 6.3.2, these will be treated separately.

6.4.2.1 Profiles of thin (2 mm) specimens

The investigation consisted of 5 series of specimens, in the longitudinal direction AL, BL and CL specimens were fatigued and MFTT profiles were made. In the transverse direction, AT and ET specimens, respectively close to the gate and far from the gate, were tested. See Fig. 6.13 for the location of the specimens in the plate. Of course the specimens that were tested were full-size specimens, as shown in Fig. 3.12. The AL type specimen is located very close to the side of the specimen, and shows no shell/core effect in the strength profile, a very small increase of the fracture strain in the core layer can be seen, Fig. 6.25.

The fatigued AL type specimens (Fig. 6.25) show a strength decrease over the entire thickness, of 10 MPa maximum and 5 MPa minimum. Also an increase in fracture strain over the entire thickness can be seen, of 4 % maximum. As there is *no core layer visible in the profile*, fibre orientation varies little over the thickness, and the fatigue damage is homogeneous over the thickness.

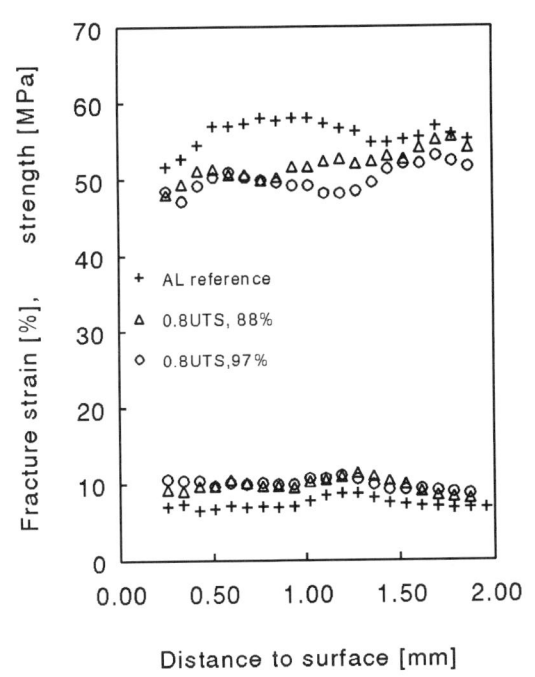

Figure 6.25   *Influence of fatigue on damage in the AL type, 2 mm thick specimens, after being fatigued at 80% of UTS, for 88 respectively 97% of predicted lifetime.*

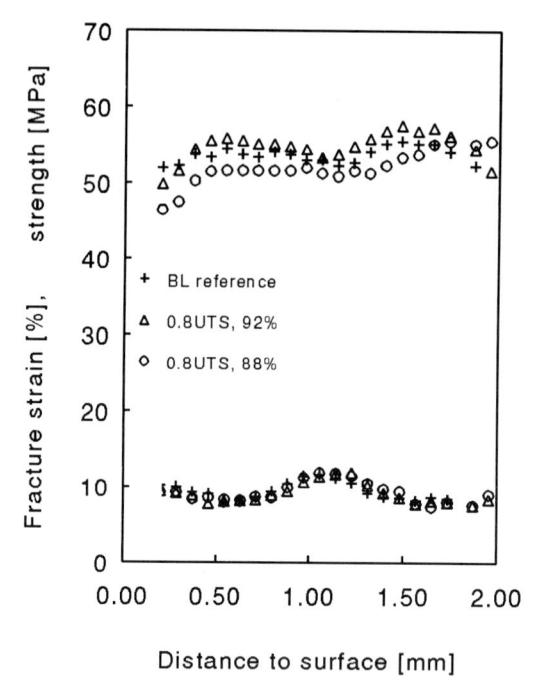

Figure 6.26 Influence of fatigue on damage in the BL type, 2 mm thick specimens, after being fatigued at 80% of UTS, for 92 respectively 88% of predicted lifetime.

For the BL type specimen (Fig. 6.26) a small effect of the core becomes visible in the strength profile, while in the fracture strain the core effect is quite pronounced. The effect of fatigueing is small, a maximum strength decrease of 5 MPa is visible. In the fracture strain no significant effect of fatigue is visible.

For the CL type specimen of Figure 6.27 the core effect is still more pronounced, in the strength as well as in the fracture strain profile. The effect of fatigue on the strength profile is similar to the two other longitudinal specimens; a strength decrease over the entire thickness of 4-7 MPa. The effect of fatigue on the fracture strain profile is more confined to the core of the specimen, although the increase of fracture strain is only 3 %, and therefore not significant.

Thus for the longitudinal specimens the effect of fatigue on fracture strain is not consistent, with fracture strain increasing over the entire thickness or over part of the thickness, or not at all. However, if a change in fracture strain exists, this is an *increase*. For all longitudinal specimens a decrease in strength over the entire thickness was found after fatigue.

The specimens in the transverse direction, perpendicular to mould flow direction, show an inverted profile, with a stronger core compared to the Shell, compare Figure 6.12. In Figs. 6.28 and 6.29 the effect of fatigue on these transverse profiles is shown. For the AT specimen, Fig. 6.28, the core effect is most obvious in the strength profile. Damage occurs over the entire thickness, though the core layer is affected most. In the fracture strain the effect is less

Chapter 6, Microfoil Tensile Tests for obtaining strength profiles

pronounced, the shell layers show no change, in the core an increase in fracture strain can be noted.

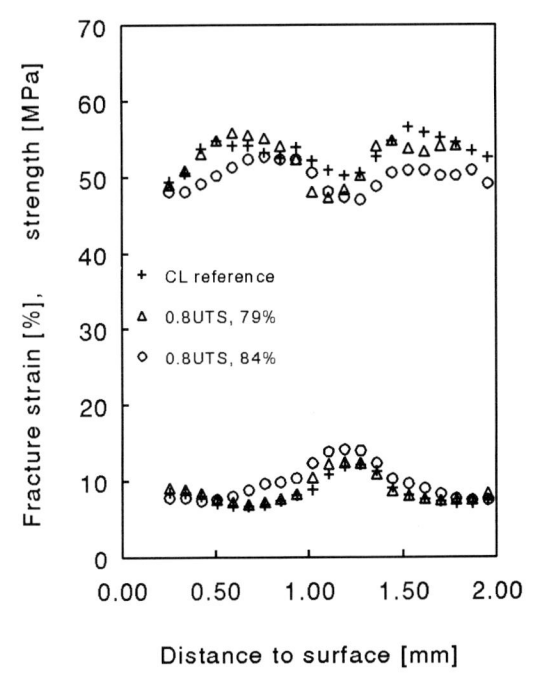

Figure 6.27   Influence of fatigue on damage in the CL type, 2 mm thick specimens, after being fatigued at 80% of UTS, for 79 respectively 84% of predicted lifetime.

The ET type specimen shown in Fig. 6.29 shows a similar strength profile as the AT type, though the core is less pronounced. The effect of fatigue on the strength profiles is similar, the core strength is affected more than that in the shell. The fracture strain shows a more pronounced core - shell pattern than the AT type specimen. After fatigue the fracture strain increases by a few percent, over the entire thickness.

Summarizing the results for the 2 mm specimens the strength is always **lowered** by the fatigue process. The entire thickness is affected similarly: damage is present throughout the thickness. This except for the case where the strong layers represent only a small percentage of the entire thickness, as in the transverse specimens. In the transverse specimens the stronger (core) layer is affected more than the weaker (shell) layers. In this case the stronger and stiffer core layer will bear most of the load. For the longitudinal specimens the stronger layers occupy more than half of the entire thickness. The strength decrease in these specimens is uniform over the thickness. The fracture strain shows less consistency, in that the fracture strain is sometimes affected over the entire thickness, sometimes the weaker layer is affected (CL type), sometimes the stronger layer (AT), and even the fracture strain can be unaffected, while the strength does decrease. This is in contradiction with the conclusions for the 5.75 mm thick CL type specimens, where it was concluded that damage is seen more easily in the fracture strain profiles (Fig. 6.24) than in the strength profiles. However, when a change in fracture strain exists, this is always an **increase**.

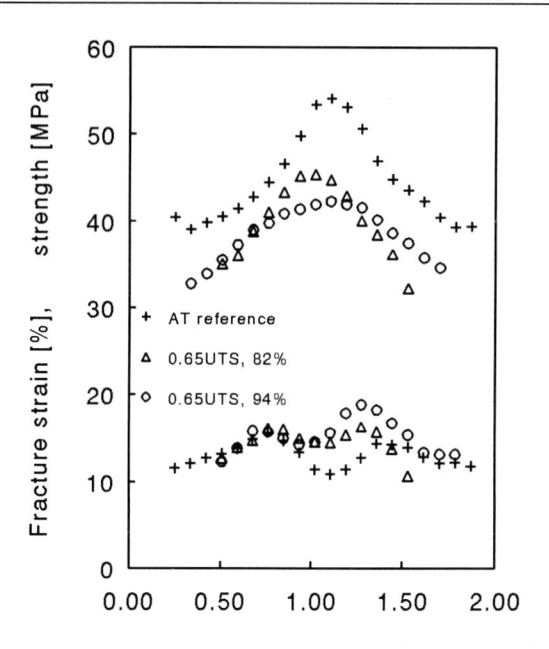

Figure 6.28 Influence of fatigue on damage in the AT type, 2 mm thick specimens, after being fatigued at 65% of UTS, for 82 respectively 94% of predicted lifetime.

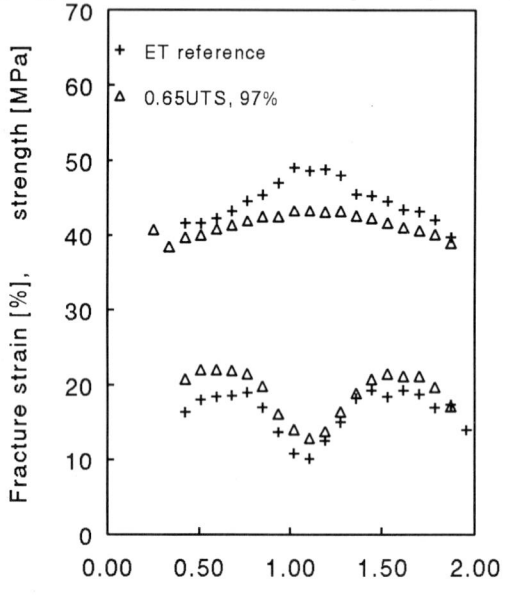

Figure 6.29 Influence of fatigue on damage in a ET type, 2 mm thick specimen, after being fatigued at 65% of UTS, for 97% of predicted lifetime.

Chapter 6, Microfoil Tensile Tests for obtaining strength profiles

6.4.2.2 Profiles of thick (5.75 mm) specimens

Apart from the CL specimens discussed in paragraph 6.4.1, profiles were made of fatigued AL and DT type specimens (see Fig. 3.12 for the location of the specimens).

The behaviour of the *AL type* specimen profiles (Fig. 6.30) with fatigue is consistent with that of the thin specimens, in that the stronger layers, the shell layers, show a decrease in strength with fatigue, while the fracture strain shows an increase. The strength and fracture strain in the core layer remain unaffected.

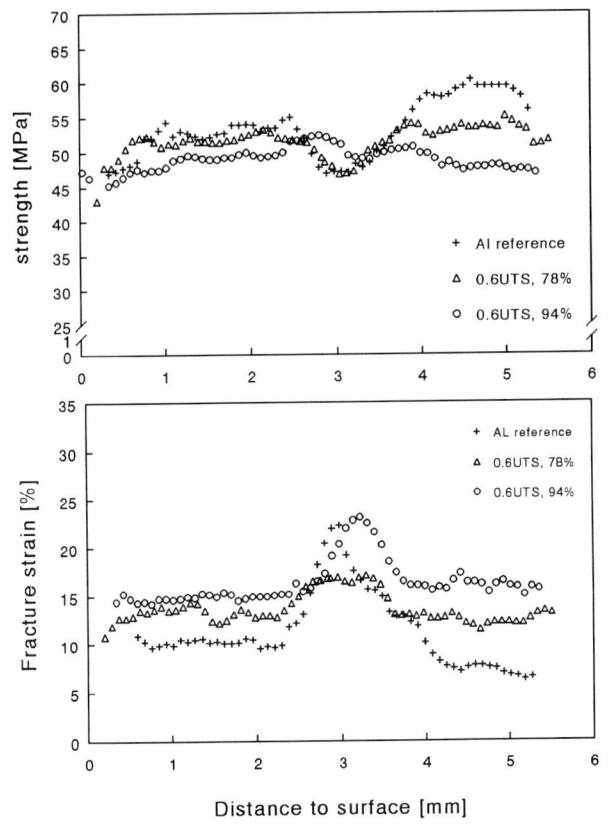

Figure 6.30   Influence of fatigue on damage in the AL type, 5.75 mm thick specimens, after being fatigued at 60% of UTS, for 78 respectively 94% of predicted lifetime.

For the *CL type* specimens a lot of information was already given in paragraph 6.4.1. Generally a decrease in fracture strength over the entire thickness was measured, accompanied by a **decrease** in fracture strain. This is opposite to what was found in all measurements described in this paragraph up to now. It is not a coincidental error in one or a few measurements, as it is seen at various load levels and percentages of predicted lifetime, Fig. 6.21, 6.22 and 6.23. This effect is also seen in Figure 6.31, where the profile for the specimen fatigued at 70% of UTS does show a general decrease in strength, and a little decrease in fracture strain as well. The specimen fatigued at 45% of UTS shows little decrease in strength, except for the foils at 3 mm distance from the surface, where stress whitening lines were visible in the foils. These foils showed an extreme local drop in strength of 10 MPa approximately, and also a drop in fracture strain.

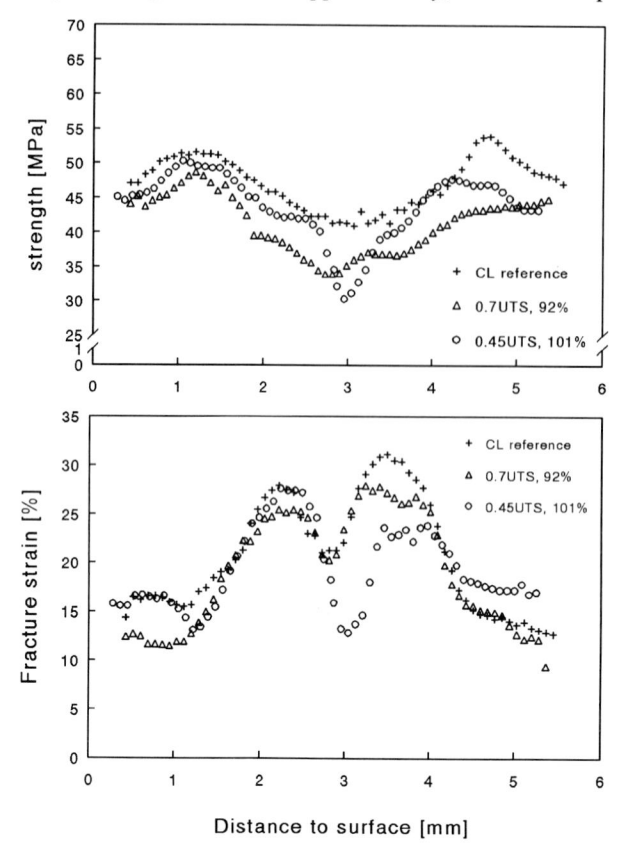

Figure 6.31   Influence of fatigue on damage in the CL type, 5.75 mm thick specimens, after being fatigued at 70 and 45% of UTS, for 92 respectively 101% of predicted lifetime.

For the *DT type* specimen only one profile of a fatigued specimen is available. This specimen was fatigued at a maximum load of 40% of UTS, and could not be broken in 5.1 million cycles. As these experiments last for several weeks, it was not possible to make a creep-speed - lifetime curve, so the predicted lifetime for this specimen is not known. In Figure 6.32 some damage is visible, especially in the shell layers, which is in accordance with results for the 2 mm specimens, and the AL type 5.75 mm specimen. In the core layer a local drop in strength is present, which was accompanied by stress whitening lines. The fracture strain is not significantly affected by fatigue.

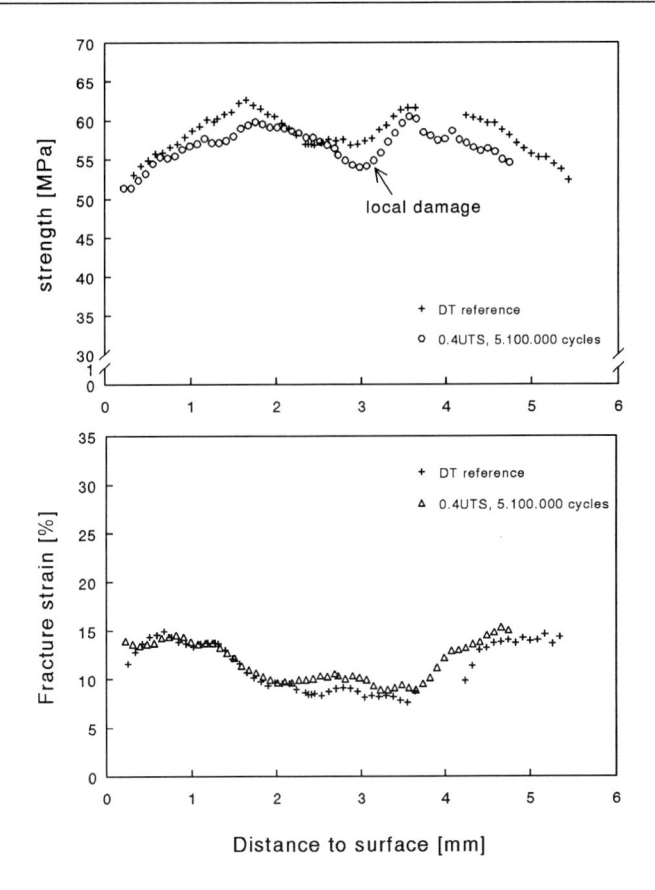

Chapter 6, Microfoil Tensile Tests for obtaining strength profiles 89

Figure 6.32 *Influence of fatigue on damage in a DT type, 5.75 mm thick specimen, after being fatigued at 40% of UTS, for 5.1 million cycles. The specimen was not broken, the expected lifetime is not known though. In the indicated area stress whitening lines were visible.*

6.4.2.3 Synopsis
In Table 6.2 an overview is given of the effect of fatigue on the MFTT profiles. For the effect of fatigue on the **strength**, results are consistent. If the specimen has been fatigued for a sufficient percentage of predicted lifetime, the strength shows a decrease, of up to 10 MPa. This strength decrease is either over the entire thickness of the specimen, or is most pronounced in the stronger layer, i.e. the shell layers in longitudinal specimens, or the core layer in the thin transverse specimens.
In some cases, exclusively at fatigue loads of 60% of UTS or lower, local decreases of fracture strength at the location of stress whitening lines were seen. This phenomenon was normally confined to the thickness of 4 - 5 foils. This *local decrease in strength* was accompanied by a *local decrease in fracture strain*. As the fracture takes place at the local damage site, the rest of the foil is not stressed to its fracture strength, and thus the fracture strain decreases.

Type		Strength decrease		Fracture strain increase		Fatigue Load
		Shell	Core	Shell	Core	
2mm	AL	5 - 10	(no core)	5	5	0.8UTS
2mm	BL	5	5	0	0	0.8UTS
2mm	CL	5	5	3	0	0.8UTS
2mm	AT	6	10	0	5	0.65UTS
2mm	ET	3	6	3	3	0.65UTS
5.75	AL	5 - 10	0	5 - 8	0	0.6/0.7UTS
5.75	CL	5 - 10	5 - 10	-5 - -15 *	-5 - -15 *	0.5-0.8UTS
5.75	DT	5	5	0	0	0.4UTS

*Table 6.2   Overview of strength decrease and fracture strain increase in the MFTT profiles after fatigue. *: Fracture strain **decrease**.*

Generally the fracture strain is seen to show an **increase**, which can be explained by the fatigue damage, which will occur first at the glassfibre - matrix interface. The debonding of the fibres makes the foils behave more like not-reinforced PolyAmide, with lower strength and higher fracture strain. The increase in *fracture strain* does not show any consistency though with the decrease in *residual strength*. In most cases a decrease in strength is accompanied by an increase in fracture strain, though sometimes the fracture strain does not show any change at all. On the other hand some specimens not showing any change in strength, did show a change in fracture strain. It is possible that two mechanisms occurring at the same time can explain this. On the one hand the debonding of the fibres causes an *increase in fracture strain*, as the material behaves more like the not-reinforced PolyAmide. On the other hand the damage (by the fatigue process, but also by the microtoming process) can be *local*. This can cause the foil to fracture at this location, before the length of the foil is stressed to the fracture strength, causing a *decrease in fracture strain* of the foil. The competition between both effects leads to the occurrence of both increases as well as decreases in fracture strain.

In contradiction to the other results, are the measurements from the CL specimens of 5.75 mm thickness. These show a strength decrease after fatigue, combined with a fracture strain **decrease**. This effect is not easily explained using the considerations mentioned above. Localisation of damage, causing the foil not to be stressed over its full length does occur, but only in some foils, which are indicated in Figs. 6.22 and 6.31. For some unknown reason fatigue damage in these foils does not result in the foils behaving more like the unreinforced polymer.

6.5 Conclusions

From the results obtained it is obvious that the MFTT method as it is discussed here, has some serious drawbacks. The foils are damaged to such a degree that the strength is seriously affected, and indeed does not even reach the strength for foils of unreinforced PolyAmide (Fig. 6.11). The fibres act more as stress concentrators, while their reinforcing effect is virtually absent. However, the fibre orientation has a strong effect on both the foil strength and fracture strain, profiles of these through thickness give a good indication of fibre orientation through the thickness of a specimen or product

The scatter is substantial in measurements of foils from different specimens, up to 4MPa has been measured (Fig. 6.9). It is much less so for foils from one specimen. Therefore it is difficult to compare strength profiles from fatigued and not fatigued, reference specimens. A decrease in strength can be only recognised if this decrease is sufficiently big. Thus the fatigue damage can be detected only above a certain minimum. Also it is difficult to quantify the strength decrease, as the maximum decrease is 10MPa, and the scatter can be in the order of 4MPa.

In spite of these drawbacks the method also shows advantages. The fact that all foils can be cut from one specimen enables the evaluation of a strength profile. Layer removal techniques can only provide a profile of stiffness. As the fatigue damage is confined to zones, the strength and fracture strain will be much more affected by fatigue than the stiffness, because stiffness is a global property of the foil, and fracture is more local. Using the MFTT method all foils for obtaining a strength profile can be cut from *one specimen*, ensuring that the *fatigue history for all foils is the same*.

The influence of the fibre orientation is similar to the situation in full-size specimens. Orientation of the fibres along the foil axis results in high strength and low fracture strain, while perpendicular fibres result in low strength and higher fracture strain. Characterizing the shell and core layers as having the fibres oriented in mould flow direction respectively perpendicular to this is too simple a representation. The degree of orientation, which is higher in the Shell layers than in the Core layer, influences the properties strongly. Consequence is that the transverse profile is not simply an inverted longitudinal profile, which can be seen in Fig. 6.12. The fact that fibres deviate from the planar orientation, especially in the core layer, causes further deviations from the expected pattern. Foil strength decreases in those layers where fibres must be cut in the microtoming process. However, the method does give a good indication of thicknesses of shell and core layers.

The foil strength is not the result of the fibre orientation alone, as it is seriously affected by the matrix properties. The decrease of foil strength towards the surface of the specimen is solely caused by a decrease in matrix strength and stiffness towards the surface, caused mainly by a higher water content at the surface. This decrease in strength towards the surface is not accompanied by any change in fracture strain.

The *effect of fatigue on the MFTT profiles* is normally a decrease in the strength and an increase in the fracture strain. This can be explained by the *debonding of the fibres* resulting from fatigue. The material behaves more like the unreinforced polymer, with a lower strength due to the lower load bearing cross-section. These changes in the MFTT profiles are more pronounced in the stronger layers, with fibres oriented in the direction of the foil axis. Because these layers have higher stiffness, they will bear most of the load during fatigue, and thus be more highly damaged. If the damage is high locally, the foil will fracture at a lower stress at this damaged site. The length of the foil will not reach its fracture stress, and therefore the fracture strain will decrease. These areas with high damage can be recognised with the bare eye as stress whitening zones, and if a foil contains such a stress whitening zone, it will generally fracture at this location.

7. Fractography

7.1 Introduction

Apart from the measurements during fatigue, a second very important method to investigate the fatigue mechanism is fractography, the macro- and microscopic observation of the fracture surfaces of the material. On a macro scale not much can be seen, the fracture surfaces in all cases appear to be brittle. It is sometimes very irregular, especially when the majority of fibres are oriented perpendicular to the fracture surface. Therefore we have to go to microscopic level, where the interactions between fibre and matrix can be observed. Scanning Electron Microscopy (SEM) is the most suitable method to analyse the fracture surfaces, especially due to the large depth of focus and high contrast.

In this chapter fractography results on glassfibre reinforced PolyAmide will be discussed. Fracture surfaces after fatigue fracture will be compared to those resulting from tensile experiments. Differences in fracture surfaces of material that was conditioned differently will be discussed, as well as the influence of the fatigue load on the fracture surface appearance. The quality of bonding between fibre and matrix influences the fatigue failure process, the effect this has on the fracture surface appearance is discussed as well.

In usual fatigue fractography, only the result of the entire fracture process can be observed. In this chapter a method will be discussed to investigate the fatigue failure process before final fracture. To accomplish this, fatigued specimens have been cryogenically broken, to reveal the structure inside the specimen.

In *literature* fractography of fibre reinforced plastics has been discussed extensively. An *overview of results* is given here, on both tensile and fatigue experiments.

7.1.1 Fractography of SFRTP in literature

Tensile experiments
In literature normally the fracture surface from a tensile experiment is reported to be microbrittle, except for composites with extremely ductile matrix material. This can be at high temperatures or high humidity (for PolyAmide), or for materials with exceptionally poor matrix-fibre bonding. Sato et al. [Sato,91] found for tensile fractured specimens a small microductile region in this microbrittle fracture surface. This microductile region is the initiation site for the crack. Sato et al. also did in-situ experiments in the SEM, revealing void formation and plastic deformation at the fibre ends, in PA6.6 containing 30%Wt. glassfibres.

Differences between fatigue and tensile experiments
SEM observations reported mainly by Lang [Lang,87b] show the following differences between fracture surfaces of glassfibre filled PA6.6 specimens broken in tensile loading and in fatigue experiments;
1 More single and multiple fibre fracture in fatigue was seen, associated with buckling or

bending of the fibres during crack closure.
2. Variations in interfacial failure site in well bonded systems; In tensile experiments the fibres remain covered with matrix material, while in fatigue the fibres generally do not show any sign of matrix material being still bonded to them. This shows the adverse effect of fatigue loading on interfacial bond strength.
3. Lang could not find any correlation between fracture surface characteristics and stress intensity difference ΔK or crack growth speed da/dn. Karger-Kocsis [Karger-Kocsis,88] reports for GFPA 6.6 matrix ductility at high da/dn, due to hysteretic heating.
4. Lang also noted a difference in matrix behaviour between stable crack growth and Fatigue Crack Propagation, with higher matrix ductility in fatigue.

Fatigue experiments
Fractography of Fatigue Crack Propagation (FCP) fracture surfaces was done by various researchers [Karbhari,90, Karger-Kocsis,90 and 88, Voss,88, Lang,87b, Lang,81, Mandell,80]. Generally the fracture surface appearance observed for the fatigue respectively the final fracture area, was similar to what was found by [Lang,87b]. Both [Karbhari,90] and [Karger - Kocsis,88] report a variation of matrix ductility over the fatigue crack growth area. Brittle-like behaviour exists at the beginning of FCP, gradually becoming more microductile with ongoing crack growth. In the *microbrittle* fatigue area very restricted pull-out was reported, in the microductile fatigue area longer pull-out lengths exist. In this *microductile* zone the FCP rate decreases. [Karbhari,90] mentions the following mechanisms: matrix pulling, fibre pull-out, crazing, shearing, fibre induced matrix damage.

Lang reports a dependence of the matrix ductility in FCP on fibre orientation, with more matrix ductility in the case of fibres perpendicular to the force. Also the fibre - matrix adhesion influences matrix ductility. A better adhesion constraints the lateral contraction of the matrix, leading to a local stress component perpendicular to the main stress direction. This enhances the severity of the local stress field, resulting in less matrix plastic deformation.

Remarkable differences exist in the degree of matrix ductility reported by different researchers. These differences are due to especially the matrix deformability, which for PolyAmide is very much dependent on the water content. Also interface strength and fibre orientation must be considered to be important parameters, in determining what failure mechanism will prevail.

For lifetime experiments very little literature is available on fracture surfaces. For carbon fibre filled PolyAmide 6, conditioned to equilibrium water content, Jinen [Jinen,89] gives some fractographic results. The matrix is highly plastically deformed. This in contrast to a fracture surface of a static creep experiment, where matrix deformation is present, but to a much lesser extent.

7.2 Experimental

The fatigue experiments were carried out as described in Chapter 4. Tensile experiments are described in Chapter 3.
SEM micrographs of the fracture surface were made using a JEOL JSM-840A after gold coating

Chapter 7, Fractography 95

of the fracture surfaces in a Balzers SCD 040. Care was taken to not damage the specimens in the SEM, using a low accelerating voltage (10kV). For a sufficiently detailed image, an enlargement of at least 250X is needed. At lower magnifications it can not be observed reliably if the matrix is microbrittle or microductile. At an enlargement of 250X the area that is viewed is only 450 x 360 µm. Therefore for some specimens the entire fracture surface was viewed and mapped, to be able to calculate the area of the microductile and the microbrittle parts.

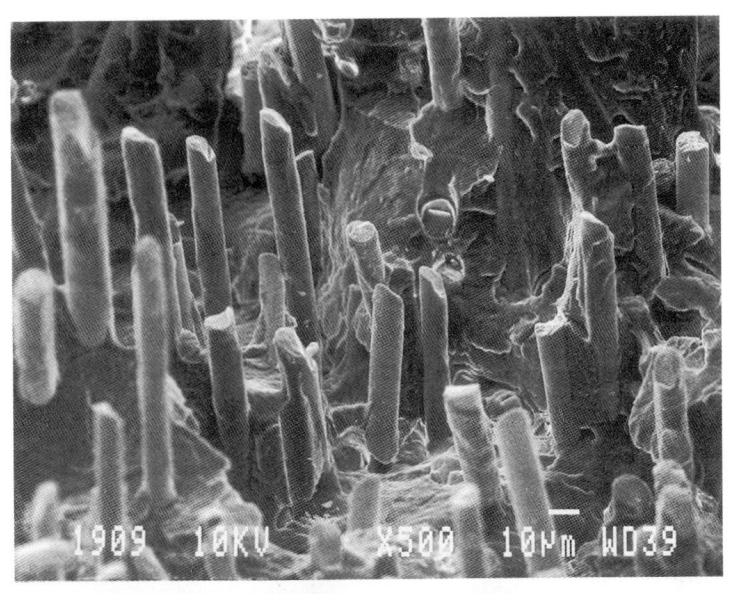

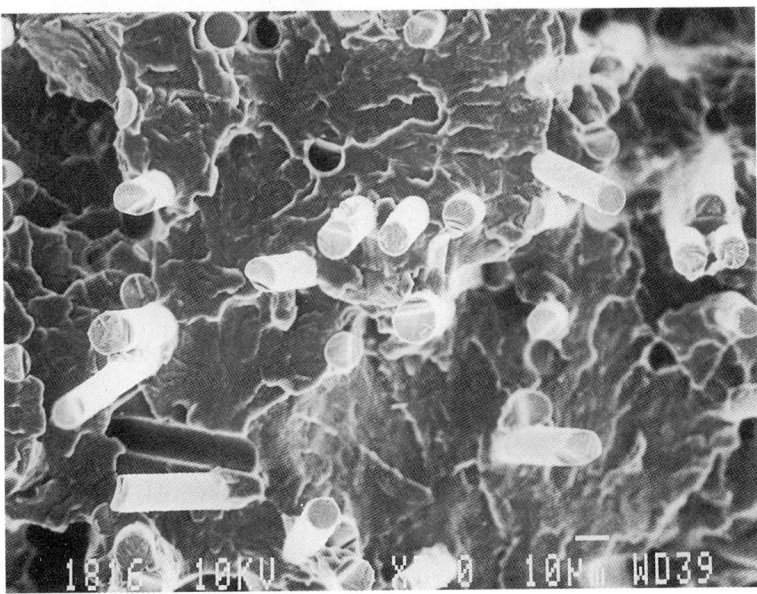

Figure 7.1   *Angle-view and top view of microbrittle fracture surface.*

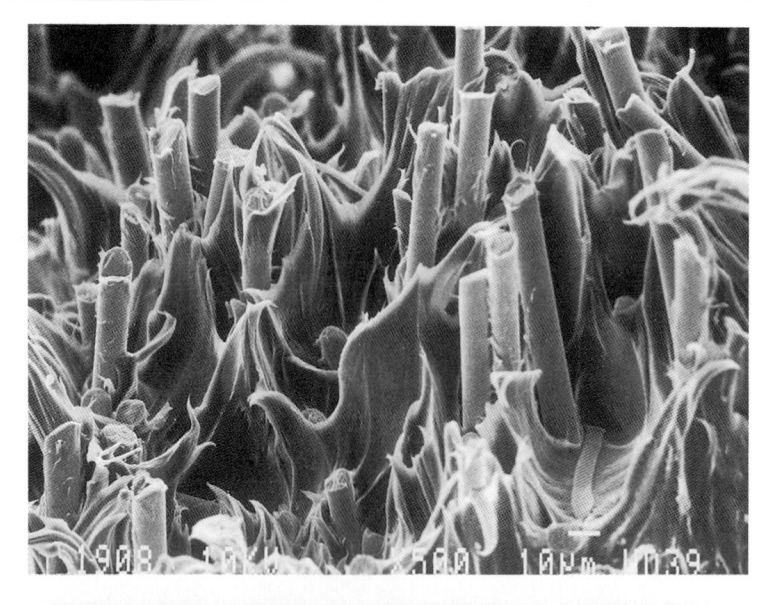

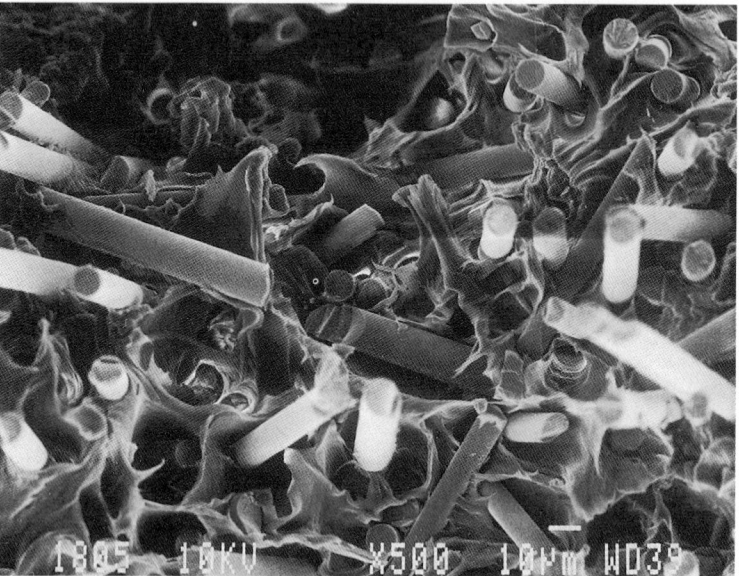

Figure 7.2   Angle-view and top view of microductile fracture surface.

To reveal the structure inside the specimen, some specimens were first fatigued for a predetermined percentage of their lifetime, and consequently fractured in three point bending after immersing them for 5 minutes in liquid nitrogen. For this investigation it was possible to use specimens that had been fatigued until fracture. The part of the specimen which was not fractured in fatigue, could be broken cryogenically. Two types of cryogenic fracture have been used. The first *parallel to the fatigue fracture surface*, the second type of cryogenic fracture was

Chapter 7, Fractography

perpendicular to the fatigue fracture surface, parallel to the specimen axis. See Fig. 7.32.

7.3 Fracture surfaces of conditioned specimens

Specimens which have been conditioned to equilibrium water content are the main interest in the research. Therefore discussion of fractographic results will begin with these. Fracture surfaces of tensile tested specimens will be discussed first, followed by those of crack growth experiments and of lifetime experiments. The terms "microbrittle" and "microductile" will be used extensively. In Figs. 7.1 and 7.2 these terms are visualised.

As can be seen in these fractographs, the main difference between these is the degree of ductility of the polymer matrix. On the microbrittle surface no ductility of the matrix can be observed, while the microductile surface shows matrix plasticity. The case for the microductile fracture surface that is shown in Fig. 7.2 is extreme. Ductility can be less. In this chapter all fracture surfaces where some ductility of the matrix can be observed (mostly seen as holes around the fibres), no matter how little, are named microductile.

7.3.1 Tensile tested specimens

In a tensile tested specimen the fracture surface is microbrittle, as in Fig. 7.1. For the majority of specimens however, a small area can be seen that is microductile, as shown in Fig. 7.3. When comparing with Fig. 7.2, the difference in degree of matrix ductility can be easily seen. The microductile area can be up to 15% of the total fracture surface. This is the area where the crack has initiated, a slow process. When this crack dominates, the crack grows much faster, not allowing the matrix material time to plastically deform.

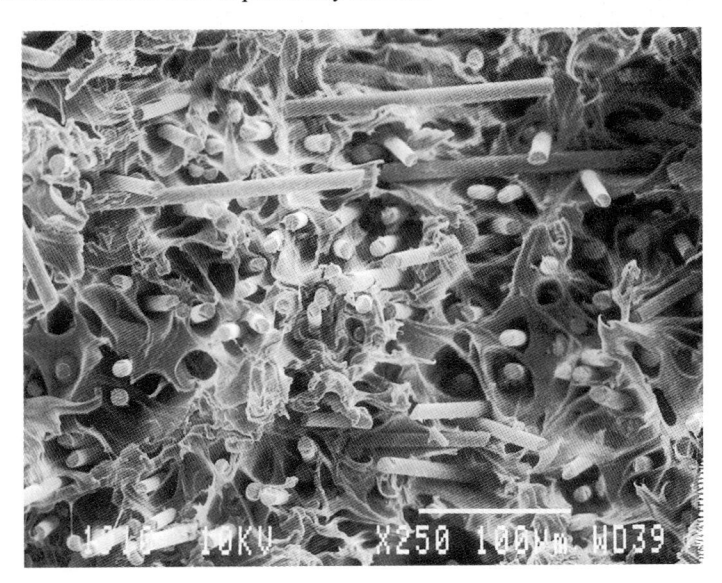

Figure 7.3   *Microductile part of tensile test fracture surface. AL type, 2mm, shell layer.*

The fibres are pulled out of the matrix. The maximum pullout length, as estimated from the fractographs, is 150 µm. Matrix material can be seen adhering to these fibres. This indicates that the fibre - matrix interface does not fail. The matrix fails, a few microns from the interface.

Generally in fatigue also a "tensile" part of the fracture surface will exist. When the fatigued area has grown to such extent, that the load bearing capacity of the specimen is lower than the maximum fatigue load, the remaining part of the specimen will break in the last load cycle of the fatigue experiment. This part of the fracture surface will be comparable with the fracture surface of a tensile test, except that the material can contain fatigue damage. As in a tensile test therefore a small microductile area of this fast final fracture may exist, while the rest will be microbrittle. This crack will generally initiate from the fatigue crack, which is also microductile: Thus part of the microductile area is associated with fatigue, while the microductile area adjacent to the microbrittle area is the beginning of the tensile crack.

7.3.2 Crack growth experiments

The fracture surface of a specimen from a crack growth experiment, see Chapter 5, was investigated. The specimen was a Longitudinal specimen (Fatigue load parallel to MFD) of 5.75 mm thickness, fatigued with 10,000 N, at 10 Hz. The specimen failed after 114.360 cycles.

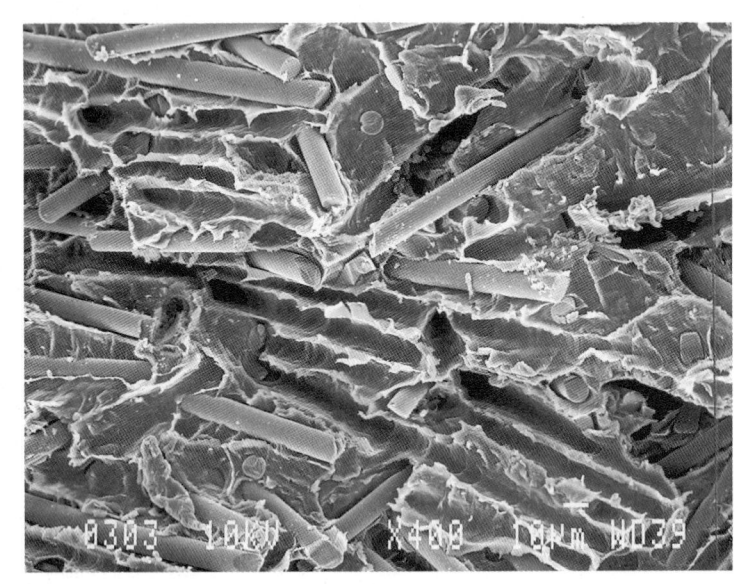

Figure 7.4   Fatigue crack growth fracture surface, close to the notch. Core layer.

From the notch (8 mm from the centre of the specimen, see Fig. 5.1) the **core** layer has a microductile appearance, see Fig. 7.4. Fibres are entirely debonded, and matrix material is highly deformed, especially at locations where fibres are close together. However, in areas containing few fibres, or fibres perpendicular to the fracture surface, the matrix deformation is less. The appearance of the core layer is microductile in the entire fatigue crack, except far from the notch (close to the side of the specimen), see Fig. 7.5. Close to the side of the specimen the orientation

Chapter 7, Fractography   99

of the fibres in the core changes from perpendicular to MFD, to parallel to MFD. However, Fig. 7.5, at 11 mm from the side of the specimen, still shows microductile behaviour. This is the beginning of the final fast fracture, which can be seen from the relatively high fibre pullout length.

In the **shell** layer the fracture from the notch starts with little ductility, Fig. 7.6. With crack growth the degree of matrix ductility increases, Fig. 7.7.

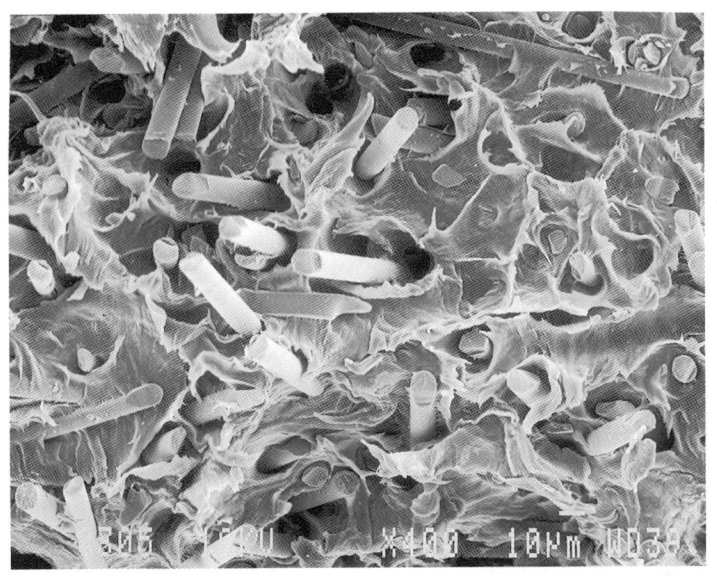

Figure 7.5   *Core area, beginning of final fast fracture.*

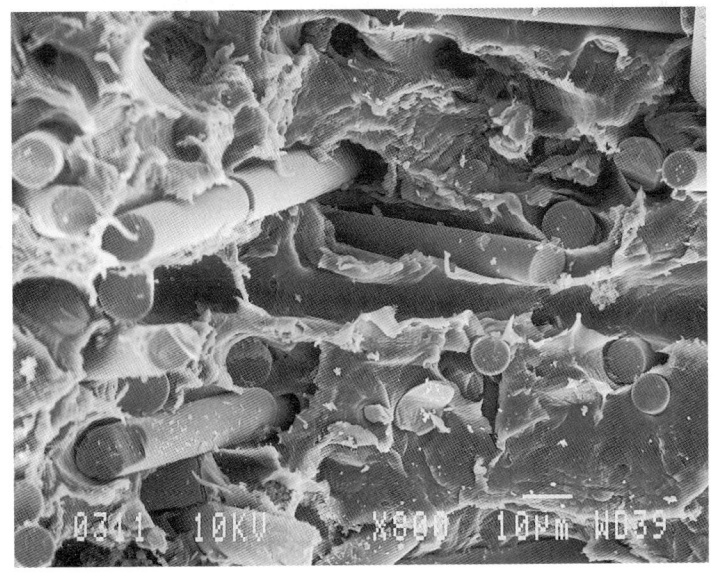

Figure 7.6   *Fatigue crack growth surface, close to notch. Shell. Note broken fibre.*

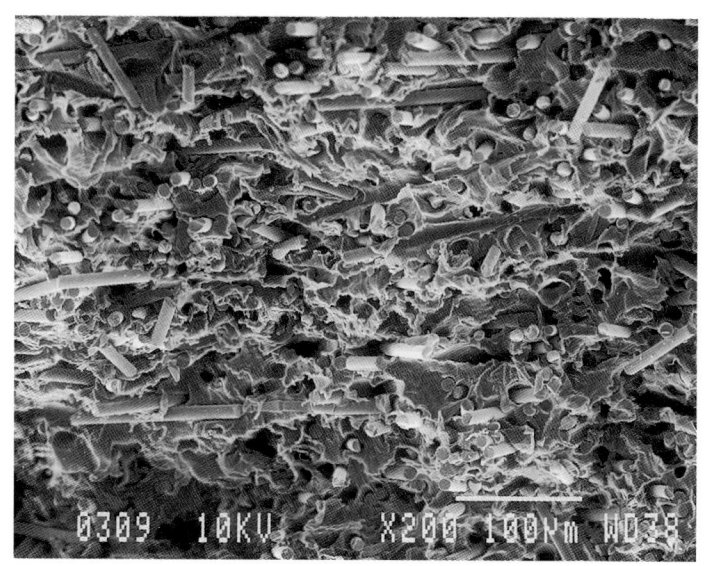

Figure 7.7   Shell area, more microductile, at 9 mm from notch.

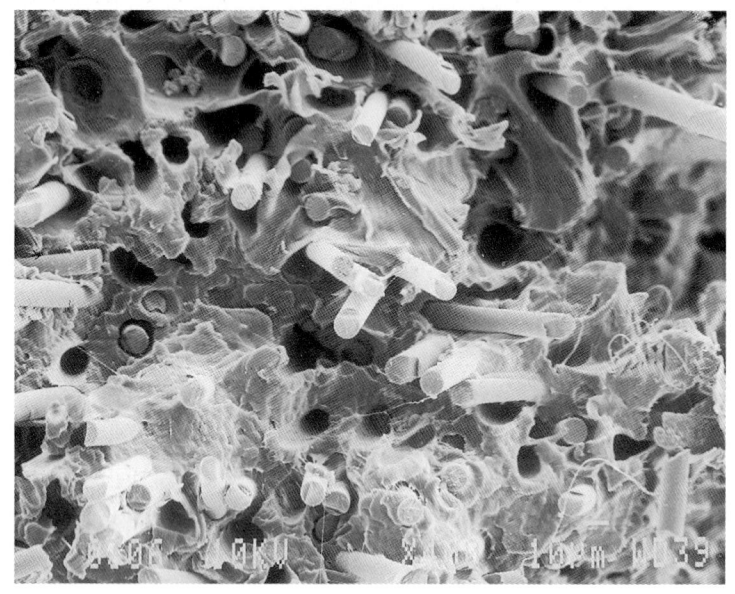

Figure 7.8   Ductile - brittle transition in the fracture surface of the crack growth experiment. Top: microductile, bottom: microbrittle.

In both the shell as well as the core, the last part of the microductile fracture surface is in fact the beginning of the fast fracture in the last load cycle. This can be concluded from the fibre pull-out length, which is shorter for the fatigue crack, compare Fig. 7.5 with Fig. 7.7. The shorter fibres in the fatigue fracture surface are caused by buckling of the fibres, during the unloading part of

the fatigue cycle [Lang,87b]. In Fig. 7.6 this effect can be seen. Some of the fibres that stick out of the fracture surface, contain a fracture. As the remaining length is too short for the fibre to break under tensile load, it must have broken under compressive load. As the fibre is not exactly perpendicular to the fracture, the compressive load will have been accompanied by a bending moment on the fibre.
As stated in 7.3.1, the remaining of the fracture surface is microbrittle, equal to the fracture surface of a tensile experiment. In Fig. 7.8 the transition in the fast fracture from microductile to microbrittle can be observed.

The appearance of the fracture surface for a crack growth experiment is as reported in literature [Karger-Kocsis,88 and Karbhari,90]. Especially in the shell layers the degree of matrix ductility increases with increasing stress intensity difference ΔK or increasing crack growth rate. The degree of matrix ductility however is not the same for the core and the shell layers. Especially in those areas where many perpendicular fibres (to loading direction) are located at short distance from each other, the degree of ductility increases.

7.3.3 Lifetime measurements

As was done for the crack growth experiments, also for the lifetime measurements the fracture surfaces were investigated. In Fig. 7.9 an overview is shown of a fatigued CL type specimen. It can be seen that in the shell layers the fracture is *extremely irregular*. In the core the fracture surface is more flat. Fig. 7.10 shows a slightly higher magnification, and some irregularity can also be seen in the core layer.

Figure 7.9   *Fatigue fracture surface, overview, CL type specimen fatigued at 0.5UTS, 154,835 cycles. 10 x 5.75 mm.*

At higher magnifications the fracture surface shows similar features as for crack growth. Fibre - matrix debonding and matrix plastic deformation in one part of the fracture surface, associated with the fatigue process, and a microbrittle part, associated with the final fast fracture. In between may be a microductile zone, being the beginning of the final fast fracture.
The degree of matrix ductility is not constant over the microductile (fatigued) part of the fracture surface. Opposite to what was observed for the fatigue crack growth fracture surface, for the

unnotched fatigued specimens the degree of matrix ductility decreases towards the ductile - brittle transition. Generally the degree of ductility is highest far from this transition, gradually decreasing until the matrix appearance is entirely microbrittle. In Fig. 7.11 a microductile fracture surface is shown, In Fig. 7.12, close to the ductile - brittle transition, the fracture is almost brittle. There are still small holes (voids) visible around the fibres though, indicating that some plasticity is present. This microductile area can be either caused by fatigue, or be the beginning of the final fast fracture, which as we have seen in §7.3.1 may also contain a microductile area.

Figure 7.10   Detail of Fig. 7.9, Irregularity of fracture surface.

There is no influence of the fatigue load level on the appearance of the microbrittle or microductile parts of the fracture surface. However, the *size* of the microductile part of the fracture surface increases with decreasing fatigue load. Explanation for this is, that the fatigue area can be larger at a lower fatigue load, before the load bearing capacity of the fatigued specimen is as low as the maximum fatigue load. In 7.4.2 the variation of the microductile area with the fatigue load level will be discussed.

Chapter 7, Fractography 103

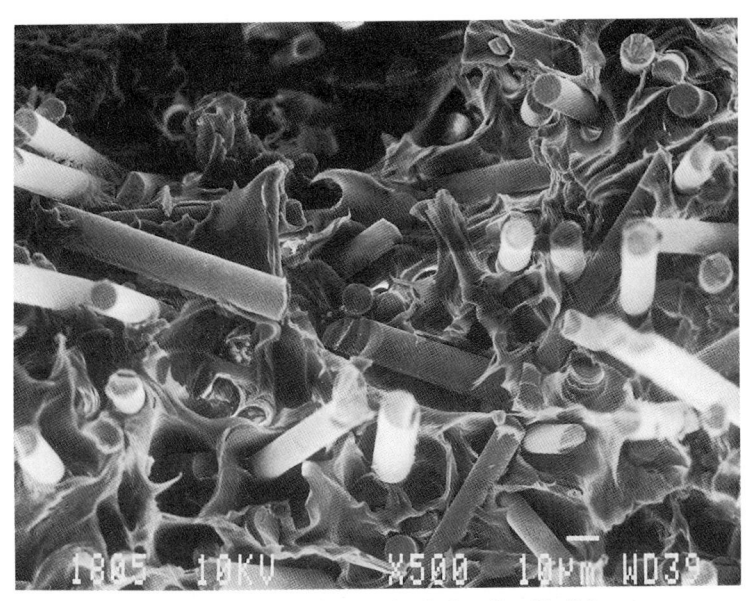

Figure 7.11   *Fatigue fracture, far from transition, high ductility. Shell layer.*

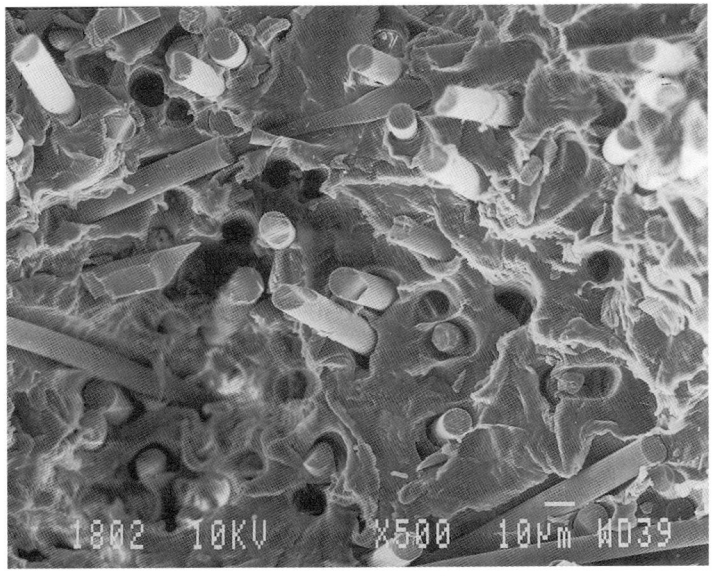

Figure 7.12   *Fatigue fracture, close to transition. Shell layer.*

Differences in fracture surfaces between different types of specimen, with a different fibre orientation distribution, are small. Appearances of the microductile and microbrittle areas is the same in all cases. There is a tendency for the microductility to occur in the layers with the fibres in the load direction. Thus in longitudinal specimens (Parallel to MFD) the ductility is more concentrated in the shell layers, while in transverse specimens the ductility occurs preferentially

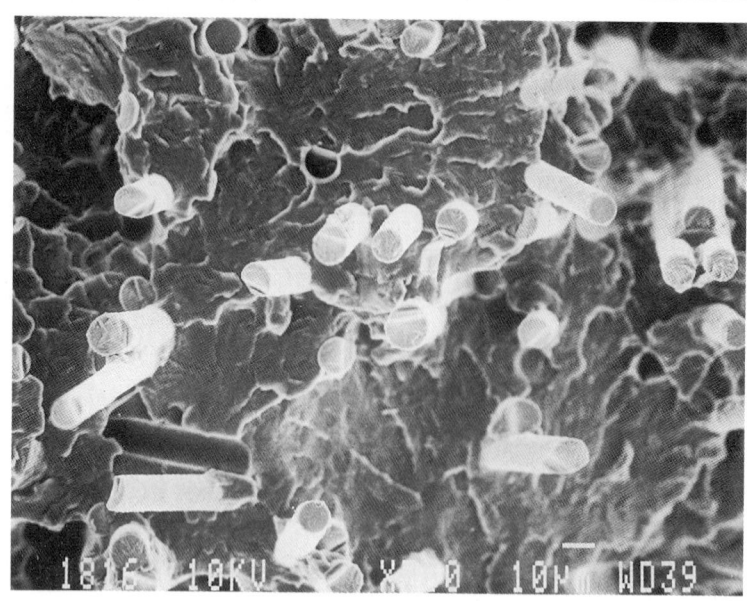

Figure 7.13 Microbrittle fracture surface of the final fast fracture of a fatigue test. AL type specimen, fatigued at 0.7UTS, 1136 cycles.

in the core layer. The fatigue damage therefore tends to be located more in the layers with fibres in the loading direction. This is probably because the fracture strain in these layers is lower and the stiffness is higher. The higher stress in these layers leads to damage, which will be augmented by the high stress intensity at the fibre end.

7.3.3.1 Comparison of fracture surfaces from fatigue and tensile experiments
For conditioned specimens some very clear differences could be observed between fracture surfaces of *fatigued* respectively *tensile tested* specimens. Figure 7.13 shows the final fast fracture for a fatigued specimen, similar to a tensile tested specimen except for the shorter pull-out length.
Matrix ductility for both cases is highly different. In case of a *tensile tested specimen* the fracture surface is microbrittle. However, a small area exists that is microductile, up to 15% of the total fracture surface. This is the area where the crack has initiated, a slow process. When this crack becomes the main crack, this crack grows much faster, not allowing the matrix material time to deform plastically. The fracture surfaces of the *fatigued specimens* contain a much larger area with microductile behaviour, and a smaller part with microbrittle behaviour.

The fibre pullout length is shorter (maximum 50 μm) in both the microductile and the microbrittle part of the fatigued fracture surface, when compared to the fracture surface for the tensile test (pullout length maximum 150 μm). Generally this is ascribed to fibre buckling in the unloading part of the load cycle, where compressive forces act on the fibre when the "crack" closes. This cannot explain the short pullout length in the microbrittle part of the fatigued fracture surface. The hypothesis for this shorter pullout length is that the crack growth speed for the final fracture area of the fatigued specimen is higher as compared to the tensile test. This can be caused by the difference in controlling of the experiments. Tensile tests are displacement controlled, with a constant cross head speed of 50 mm/min, while fatigue experiments are force controlled, cross head speed in the final fast fracture can be significantly higher than 50 mm/min. This higher speed will lead to a shorter time for the fibres to be pulled out of the matrix, and a higher tendency to fracture of the fibres.

On the fibres that are pulled out of the matrix in a tensile test, as well as the fast fracture area of a fatigue test, matrix material can be seen adhering to the fibre. In this case not the interface has failed, but the matrix material some distance from the interface. Fibres in the microductile part of a fatigue test are completely clean, with no matrix material adhering to them: In the case of fatigue the interface itself fails. We can conclude that the fatigue process has a detrimental effect on the fibre matrix bonding.

Conclusions:
- The matrix material fails to a larger extent in a ductile way in the fatigue test rather than in the tensile test. For the microbrittle areas no differences could be observed (apart from the difference in pull-out length).
- The pull-out length in the fast fracture (microbrittle) area is longer in the case of a tensile test, as compared to a fatigue test.
- The fibre - matrix bonding remains intact in the microbrittle areas, while in the microductile areas debonding occurs.
- The fibre is less debonded from the matrix in the microductile area in the case of a tensile test. This indicates that not only the different stress state due to the plasticity plays a role in debonding, but also the cyclic load itself.

7.3.3.2 Broken fibres initiating matrix failure

One particular phenomenon that was observed regularly, and that was not reported before in literature, is the occurrence of fibre fracture in tension. Fibre fracture during the unloading stage due to bending or buckling [Lang,87b] is reported frequently in literature, and could be confirmed in this investigation, see Fig. 7.6. However, shown in Figs. 7.14 and 7.15 are two typical examples of the *tensile fracture* of glassfibres: A fractured fibre surrounded by a circular matrix crack. The phenomenon was observed *only in the microductile to microbrittle transition area*, in both fatigued and tensile tested specimens. That it is actually a fibre broken in the fatigue process could be confirmed by comparing both fracture surfaces: On both surfaces the broken fibres can be matched. It is a rare phenomenon, in average about 1 in 50 fibres in this transition area is broken in tension. This is caused by the relatively short fibres. Only a very limited number of fibres are longer than the critical length. Of those fibres that are long enough for the fibre to be fully loaded, only a limited number has both fibre ends at a distance from the main damage zone, long enough for neither end to be pulled out. When looking closely at the broken fibre and surrounding matrix (Fig. 7.15) a concentric matrix crack is noticed. The observed matrix cracks growing from a fibre fracture, are very similar to those presented by ten Busschen and Selvadurai [Busschen,95, Selvadurai,95]. In this pair of articles both experimental investigations and computational modelling are compared for matrix fracture initiating at a fibre fracture in a single fibre fragmentation test. Either penny shaped cracks, conical cracks or combinations of those were found in experiments and could be modelled as well using micromechanics. Condition for these cracks to occur is a good fibre - matrix bonding.

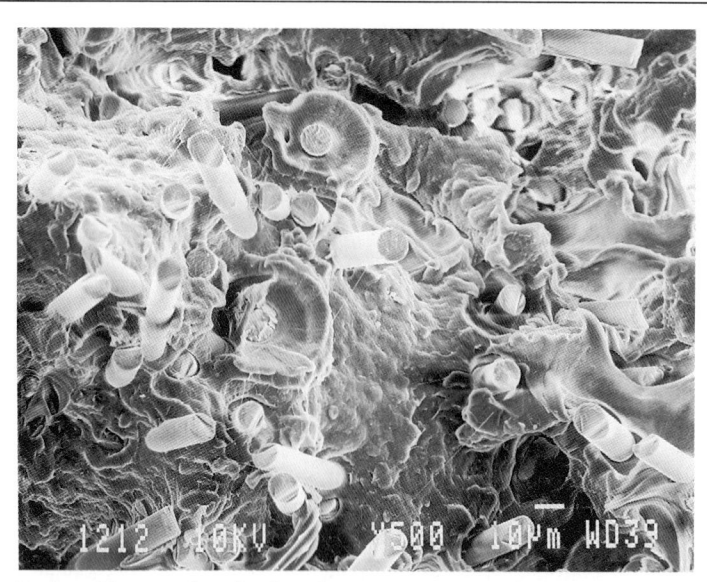

106　　　　　Influence of fibre orientation on fatigue of short glassfibre reinforced PolyAmide

Figure 7.14　　*Broken fibres in ductile - brittle transition zone. Shell layer.*

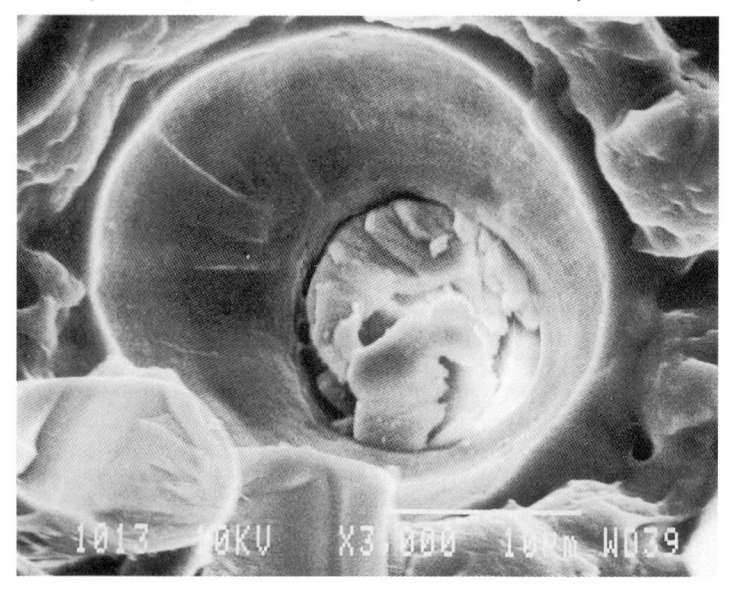

Figure 7.15　　*Close-up of broken fibre. Shell layer. Broken fibre in the centre.*

The reason why the material cracks, is the high stress intensity, due to the presence of the very stiff fibre close to the matrix. This causes stable crack growth in the polymer emanating from the broken fibre. The stress intensity decreases, when going further from the fibre. Therefore it is possible that the stress intensity drops enough for the crack to stop. This explains the observations in the Figs. 7.14 and 7.15: The material in front of the crack or damaged area is highly strained, high enough for a long fibre in this area to break. At this moment the matrix in this area is still intact. At the fibre fracture a very high stress intensity in the matrix exists, and causes a crack to initiate and grow, from the fibre fracture. When the crack grows, the stress intensity lowers, and the crack may become arrested. This pattern is consequently revealed when the main crack grows and surpasses this circular crack. The similarity between the patterns found here and the ones found by ten Busschen and Selvadurai in single fibre specimens with a Polyester resin matrix is quite surprising, taking into account the different circumstances. In Fig. 7.14 it is visible that the crack around the fibre is not always completely circular, depending on the presence of other fibres.

7.4 Influence of conditioning

Investigations were initiated using conditioned samples, as most parts in service will generally be exposed to an atmosphere with humidity between 30 and 70%. Large differences were found between these results using conditioned materials and results in literature, where mostly dry as moulded material was investigated. The material of course can also be used in conditions, where the PolyAmide can be entirely saturated with water. Therefore it was decided to compare the three differently conditioned materials. Fracture surfaces of dry as moulded and 100% wet specimens are compared with the results for conditioned specimens discussed before. The results for the fatigue performance of these differently conditioned materials are given in §4.5

7.4.1 Appearance of fracture surface

Dry as moulded material
Figs. 7.16 and 7.17 show details of the fracture surface of a fatigued dry specimen, which can be compared to Figs. 7.11 and 7.12 for the conditioned material. Figs. 7.16 and 7.17 show the microductile area, where the fatigue crack or damaged zone has developed. The ductility is much less, compared to conditioned specimens. In tensile experiments no microductility was found, opposite to the small microductile area found in conditioned material. In Figure 7.18 a schematic representation is shown of both dry as moulded as well as conditioned fatigue fracture surfaces. For comparison also the case for a tensile test is shown. In between the fibres, the matrix shows some ductility, but hardly any close to the fibres. The fibre - matrix bonding is much better compared to conditioned material. This shows in the fact that almost no holes around the fibres are visible. Obviously from the photographs the absorbed water has a considerable influence on both matrix ductility and bonding. The quality of the bonding also influences the ductility of the matrix, as a strong bond puts a high constraint on the matrix [Lang,87b].
No differences in appearance of the fast fracture (microbrittle) part of the fracture surfaces can be seen.

Contrary to the conditioned specimens, the dry specimens show a lower ductility at the beginning of the crack. This was also seen in FCP, see §7.3.2. This almost microbrittle fatigue surface is shown in Fig. 7.16. With crack growth the ductility increases slightly (Fig. 7.17), and then embrittles again towards the ductile-brittle transition. For the conditioned specimens this smaller ductility at the beginning of "crack" growth was not observed, only a decrease in ductility towards the ductile-brittle transition can be observed.

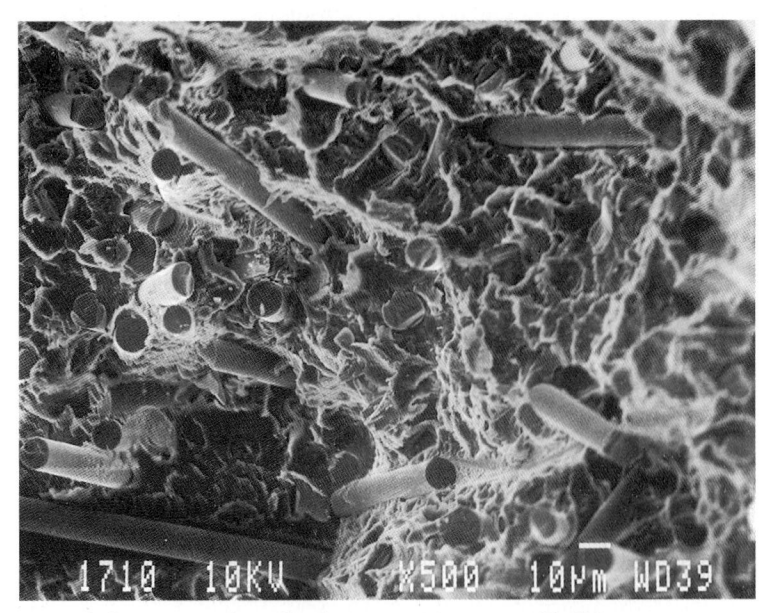

Figure 7.16 Fatigue fracture surface, far from transition zone. Shell layer.

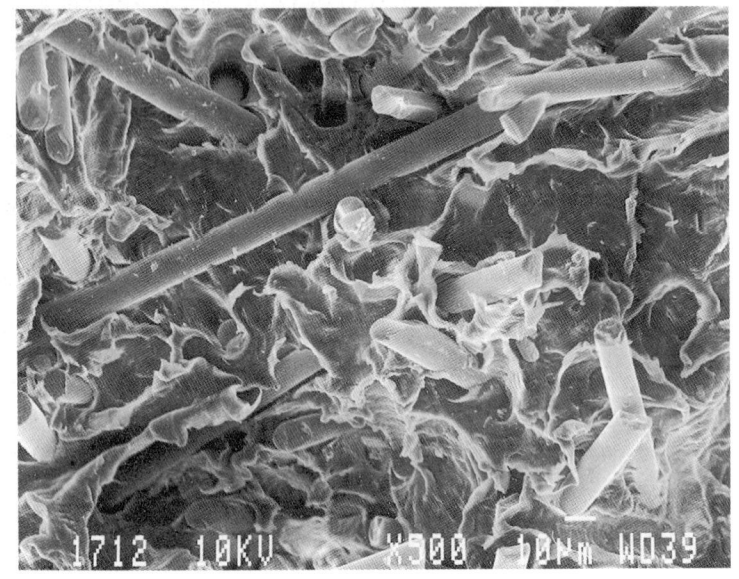

Figure 7.17 Fatigue fracture surface, close to transition zone. Same specimen as Fig.7.16 Shell layer.

Comparing the *dry as moulded* to the *conditioned* specimens (Figs. 7.16 and 7.11 respectively), it is seen that in the first case all fibres are broken at the fracture surface. In the second case some fibres are pulled out of the deformed surface. In literature this effect of broken fibres is attributed

Chapter 7, Fractography

to fibre buckling during the unloading stage of the load cycle [Lang,87b]. The lower ductility of the matrix and the greater constraint on the matrix through better bonding in the dry material, will result in a higher susceptibility of the fibres to buckling. In the conditioned material the ductility of the matrix material leaves more space for the fibres to move during unloading, leading to less fibre breakage.

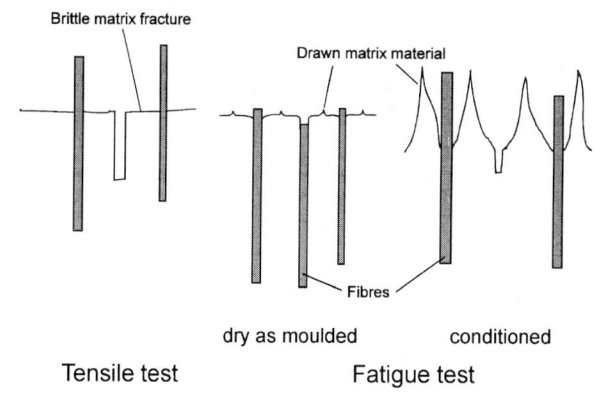

Figure 7.18   Schematic representation of the fracture surfaces of tensile experiments, and of dry as moulded and conditioned specimens that were fatigue tested.

Figure 7.19   Microductile part of tensile tested, 100% wet AL type specimen. Shell layer, 500x.

100% wet material
Contrary to the dry as moulded material which shows relatively microbrittle fracture surfaces, the case for 100% wet material is entirely opposite. Compared to conditioned material the matrix deformation is much higher, even for a tensile experiment. For a tensile experiment more than

110   Influence of fibre orientation on fatigue of short glassfibre reinforced PolyAmide

half of the fracture surface is microductile, as shown in Fig. 7.19. The matrix material also is more plastically deformed, compare with Fig. 7.3.
For the fatigue experiments the same counts, the degree of matrix ductility is much higher compared with the conditioned material. Figs. 7.20 and 7.21 show this extreme ductility. This

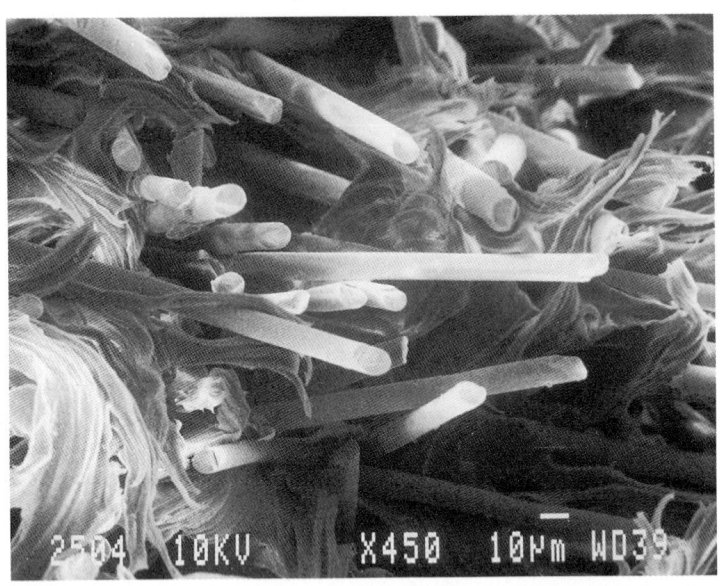

Figure 7.20   Microductile fracture surface. AL type specimen, 100% wet. Fatigued at 0.5UTS, Shell layer.

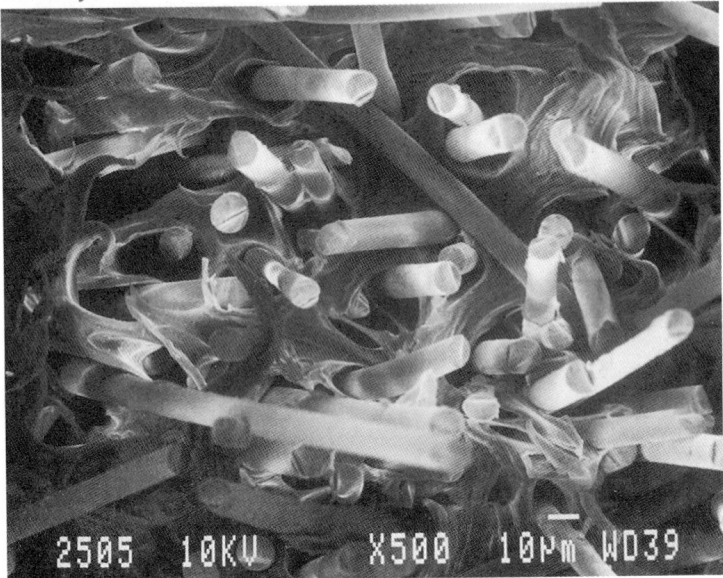

Figure 7.21   Microductile fracture. Less ductility compared. to Fig.7.20 (same specimen), Shell layer.

Chapter 7, Fractography   111

is caused on the one hand by the plasticising effect water has on the PolyAmide matrix, on the other hand by the detrimental effect of water on the interface strength. On the fatigue fracture surfaces of conditioned material a microbrittle part is always visible, for 100% wet material, this is the case for high fatigue load levels only.

Apart from the higher degree of matrix ductility, the only difference visible on the fracture surface is the bigger pullout length of the fibres, compare Figs. 7.20 and 7.21 with Figs. 7.11 and 7.12. Explanation for this is similar as for the different pullout length between dry as moulded and conditioned material: The higher matrix ductility provides more space for the fibres to move during the unloading part of the fatigue cycle, leading to less fibre fracture due to buckling.

7.4.2 Relation between microductile area of fracture surface and load level

As was shown, generally part of the fracture surface of a fatigued specimen is microductile, and part is microbrittle. In most cases one microbrittle and one microductile part exist, only in a few fracture surfaces, apart from the main microductile area, a small secondary microductile area is observed. For a number of specimens the relative sizes of the microductile and microbrittle parts were measured. This was done by mapping the position of the ductile - brittle transition on the fracture surface, see Fig. 7.22 , and calculating the respective areas. The SEM has to be used to recognise the location of the transition. Still the transition tends to be so gradual, especially for the case of the dry as moulded and 100% wet specimens, that the position of the transition can not be determined precisely. In these cases a uncertainty of maximum 0.3 mm has to be counted

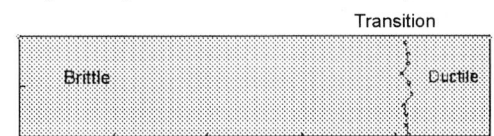

Figure 7.22   *Map of a fracture surface, indicating microbrittle and microductile areas, and the location of the transition.*

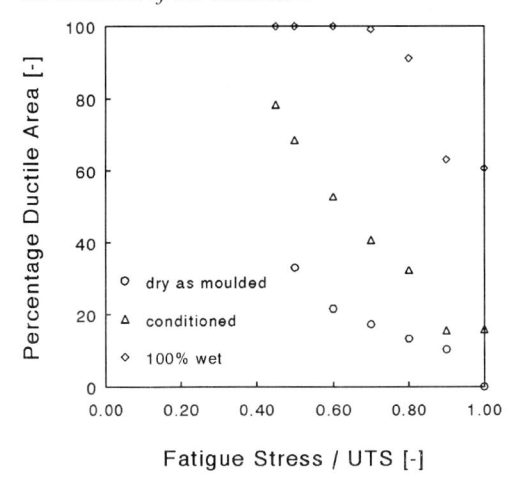

Figure 7.23   *Percentage microductile area versus normalised fatigue stress.*

with, while for the conditioned specimens this can be 0.1 mm or even less. However, the width of the specimen is 10 mm, so the maximum error in the measurement of the microductile area is below 5%.

Results are shown for differently conditioned materials in Fig. 7.23. The percentage microductile area is plotted versus normalised fatigue stress. The value indicated at normalised fatigue stress = 1 is the percentage microductile area for the tensile experiment. For each class of differently conditioned material the microductile area decreases when the fatigue stress increases. The differences between differently conditioned specimens are even more obvious: Dry as moulded material shows little ductility, while the 100% wet specimens are almost entirely microductile. The conditioned specimens show a degree of microductility in between.

7.4.2.1 Modelling

The results can be represented physically by using *fracture mechanics*. By assuming that the microductile area is a crack, the values measured can be compared with calculated values for the residual stress of a specimen containing such a crack. The geometry used is shown in Fig. 7.24. a is the crack depth, while W is the width of the specimen. The crack front is assumed to be straight, which is not always the case with the actual measurements of the ductile - brittle transition. However, this is a reasonable approximation of reality.

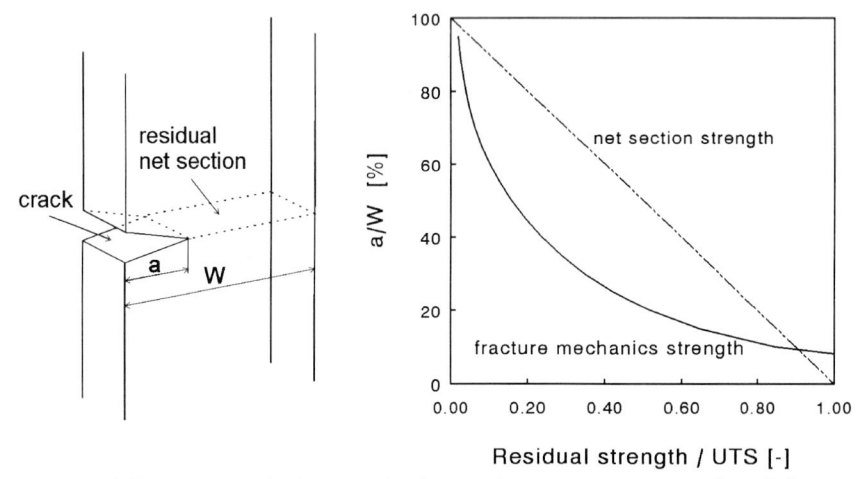

Figure 7.24   *Modelling the residual strength of a crack, net section strength and fracture mechanics strength.*

The simplest case is to assume that the residual strength is based on the net section. The reduction in net section leads to a proportional reduction in residual strength:

$$\sigma_c = UTS \cdot (1 - a/W) \qquad 7.1$$

The residual strength for this case is represented by the straight line in Fig. 7.24.

Linear Fracture Mechanics gives the curved line in Fig. 7.24. If the material is considered to be notch sensitive, instable crack extension occurs when the stress intensity K_I reaches the critical

Chapter 7, Fractography 113

value K_{Ic}.

$$K_I = \sigma_{nom} \cdot Y \sqrt{\pi a} \qquad 7.2$$

Where σ_{nom} is the nominal stress, a is the crack depth and Y is a geometry factor:

$$Y = 1.12 - 0.231(a/W) + 10.55\,(a/W)^2 - 21.72\,(a/W)^3 + 30.39\,(a/W)^4. \qquad 7.3$$

W is the width of the specimen (10 mm). The residual strength σ_c at fracture is calculated by assuming the stress intensity to be equal to the critical stress intensity: $K_I = K_{Ic}$.
Values for K_{Ic} were found in literature [Malzahn,84], and varied from 7 to 11 Mpa√m, for dry as moulded material. $K_{Ic} = 9$ MPa√m was used in the calculations.

7.4.2.2 Comparison of results with modelling

When the measurements and calculated values are compared, Fig. 7.25, the values for the *dry as moulded material* are close to the fracture mechanics curve. The measured values for the ductile percentage of the fracture surface for the *conditioned specimens* fit a straight line, parallel to the line for the net section strength. This line goes through the percentage ductile area for a tensile test (15-16% in this case). This indicates that the beginning of the final fracture is also a relatively slow process, enabling the matrix material to deform plastically. 15 - 16% of the fracture surface, which is microductile, is not to be associated with the fatigue crack, but with the beginning of the final fast fracture. For the *100% wet material* shifting of the straight (net section strength) line to the value for the microductile part of the tensile experiment, also fits the results, allowing for some scatter.

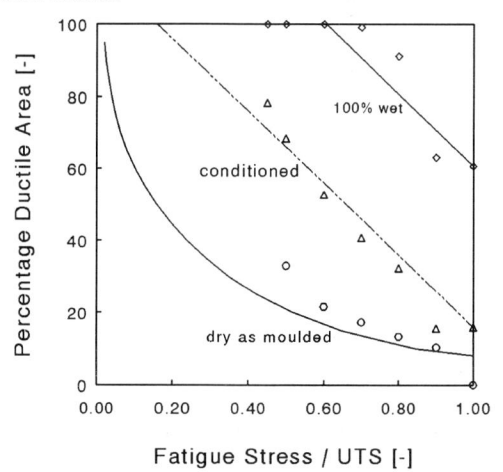

Figure 7.25   Comparison of measurements of percentage microductile fracture surface and calculated residual strengths.

The measurements for the dry as moulded specimens are close to the curved line, the notch sensitive case, while the conditioned specimens are close to the straight line for the net section strength. So the conditioned material is entirely notch unsensitive, while the dry as moulded material is notch sensitive. This result can also be interpreted in another way, the dry as moulded fatigue damage is a real crack. In the conditioned material a bridged crack develops with fatigue. The bridges that connect both "fracture surfaces" prevent the final fast fracture until the moment when the cross sectional area is reduced to the size where the stress in this area equals the tensile strength (UTS) of the material. This can be an explanation also for the results for the wet material. Bridges connect the crack walls, thus the crack can transfer some of the fatigue load, and the "crack" can grow to a size bigger than what would be predicted from the load carrying ability of the residual net section.

7.5 Influence of improved fibre coating

Dry as moulded material
For both 2mm and 5.75 mm thick dry as moulded specimens, fractographs have been made of specimens fatigued at 45%, 50% and 70% of UTS. Representative examples are given in Figs. 7.26 - 7.29. For all load levels the experimental grade shows better bonding compared to the standard grade. This shows in the fact that for the experimental grade no space is visible between the fibre and matrix (See detail Fig. 7.30), as is the case for the standard grade. In Fig. 7.31 the difference between both grades in terms of bonding is made visible, concluding from the photographs. Due to this better bonding, the matrix material in the experimental grade shows a more brittle-like behaviour, while the standard grade shows higher ductility. This is the consequence of the higher constraint on the matrix material in the experimental grade. When comparing the experiments executed at 50% of UTS to the ones executed at 70%, the higher load for both grades shows less fibre - matrix bonding, and higher matrix ductility.

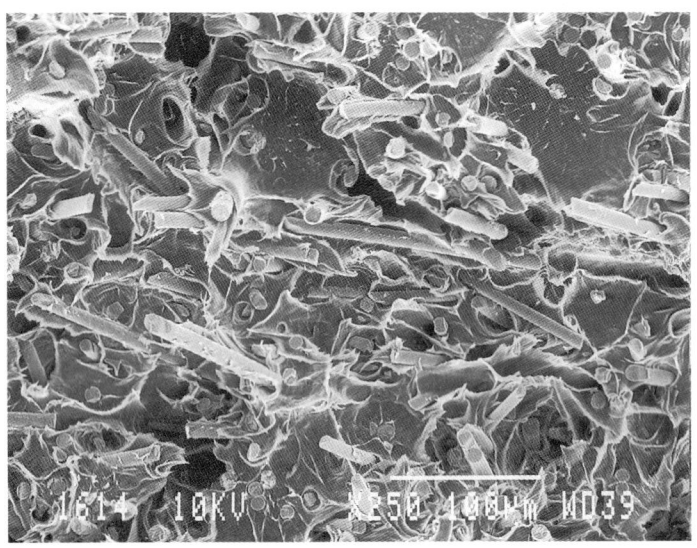

Figure 7.26   *Microductile fracture surface. AL type, dry, experimental grade. Fatigued at 0.5UTS. Shell layer. 250x.*

Chapter 7, Fractography 115

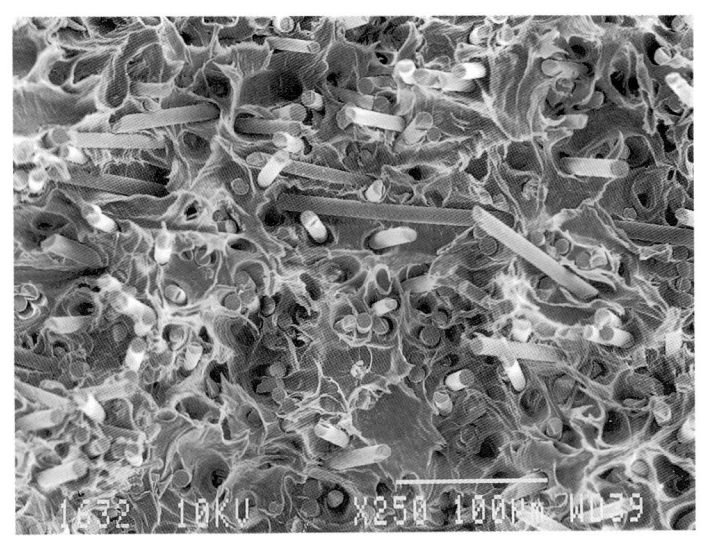

Figure 7.27 Microductile fracture. AL type, dry, standard grade. Fatigued at 0.5UTS. Shell layer. 250x.

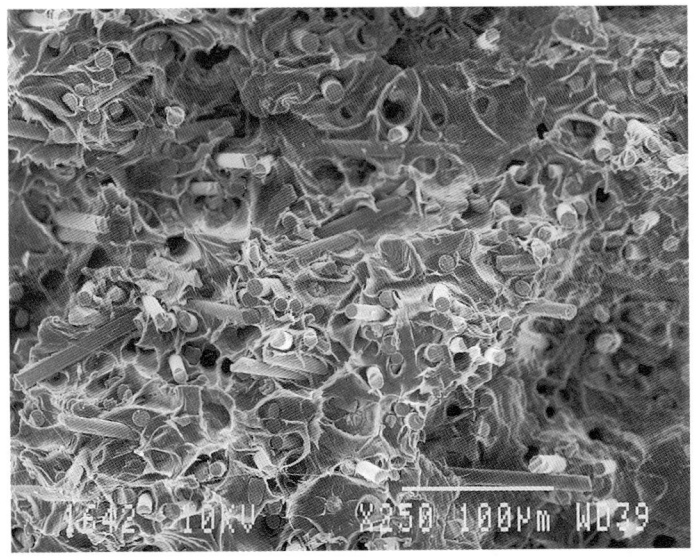

Figure 7.28 Microductile fracture surface. AL type, dry, experimental grade. Fatigued at 0.7UTS. Shell layer.

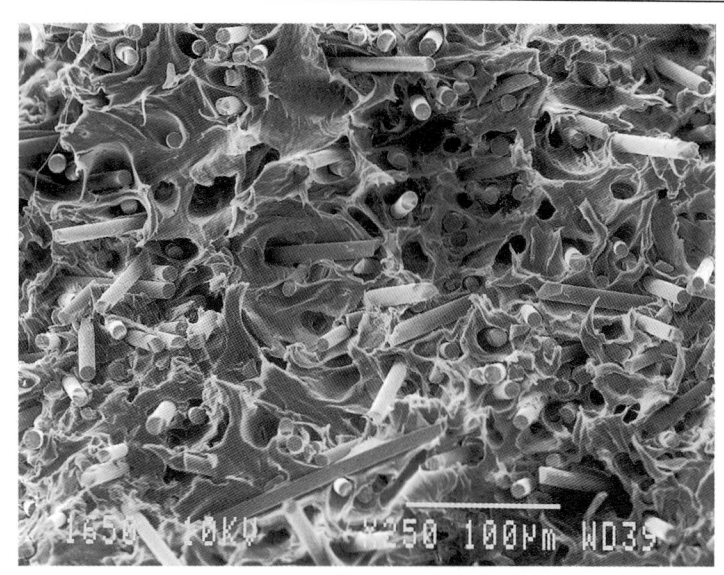

Figure 7.29   *Microductile fracture. AL type, dry, standard grade. Fatigued at 0.7UTS, Shell layer.*

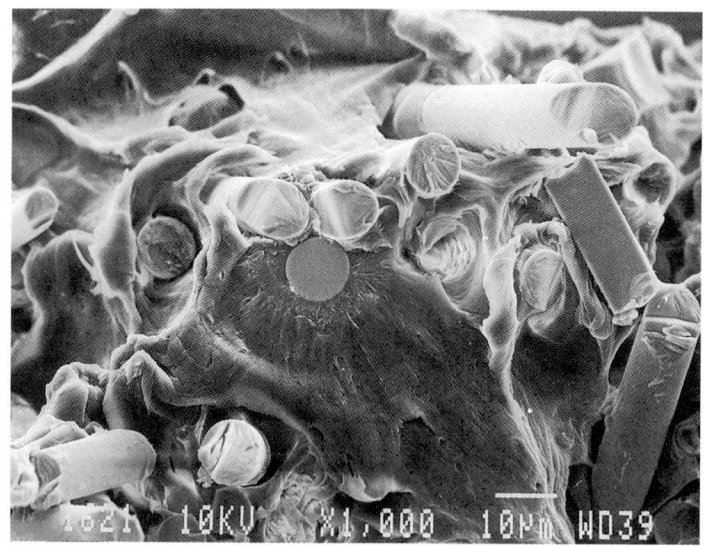

Figure 7.30   *Detail of experimental grade fracture surface, dry as moulded, fatigued at 0.5UTS. Microductile area. Practically no holes visible around the fibres.*

Conditioned material
For the conditioned materials less differences could be observed, between standard and experimental grades. Matrix ductility seems to be the same (for the ductile, fatigue, part of the fracture surface), the bonding of the matrix material to the fibres is better for the experimental

Chapter 7, Fractography 117

grade, which can be concluded from the holes around the fibres which can be seen in the standard grade, and are nearly absent in the case of the experimental grade material [Horst,96b].

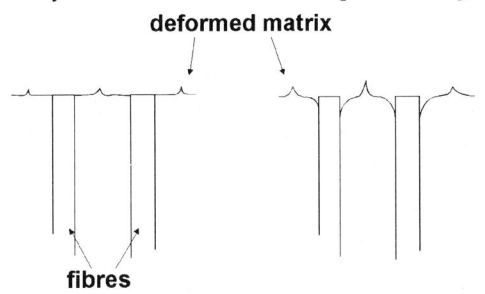

Figure 7.31   Comparison of experimental (left) and standard (right) grades. Visualisation of fracture surface for dry as moulded material.

Size of the microductile zone
According to the measurements presented in 7.4.2, also for material containing experimental and standard fibres, the relative sizes of the microductile area were measured. This was done both for dry as moulded, as well as for conditioned specimens. No significant differences could be found however, between both grades of reinforced material [Horst,96b].

Concluding: the experimental grade shows better matrix bonding compared to the standard grade. From SEM photographs it was noticed that: more matrix material is adhering to the fibres, less holes exist around the fibres and less matrix ductility is present. For load levels of 50% of UTS and lower, and in dry as moulded material, the experimental grade shows no debonding at all, and an almost brittle appearance of the matrix material in between the fibres. The sizes of the microductile area on the fracture surfaces showed no difference between the two grades.

7.6   Cryogenically broken pre-fatigued specimens

Fractography is of course very useful, but has the drawback that the occurrences during fatigue may be obscured by changes that occur later in the fatigue process. To be able to observe closely not solely the last events in the fracture process, but the actual occurrences during fatigue, the following method has been used to reveal the structures inside the material. Specimens were fatigued for a predetermined percentage of their expected lifetime. Fatigue lifetime can be accurately predicted using the creep speed method, §4.3.3. Consequently the specimens were immersed in liquid nitrogen, and cryogenically broken in three point bending. Two methods were used for this, as shown in Fig. 7.32. Specimens were broken both transverse and longitudinal to the specimen axis. In the first case the cryogenic fracture is parallel to the fatigue fracture, in the second case the cryogenic fracture is perpendicular to the fatigue fracture, parallel to the loading direction. This method can be used to observe structures caused by fatigue, because all ductile behaviour visible on the cryogenic fracture surface must be caused by the fatigue process, as the cryogenic fracture gives pure brittle behaviour. This was proven using not-fatigued specimens, which were cryogenically broken and show microbrittle behaviour over the entire fracture surface.

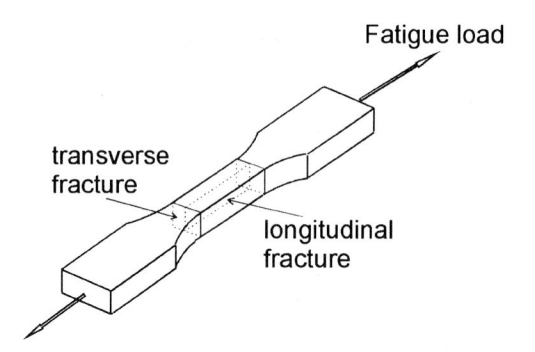

Figure 7.32   Explanation of transverse and longitudinal fractures, that were executed cryogenically.

7.6.1   Transverse cryogenic fracture

Figs. 7.33 and 7.34 show representative examples of cryogenically broken samples, that were fatigued first. The samples were fatigued at 70% of UTS for 350 cycles, approximately 90% of the expected fatigue lifetime. In Fig. 7.33 it is clearly seen how, although the fracture of the matrix is brittle, voids exist around the fibres. Almost no matrix material can be seen adhering to these fibres.

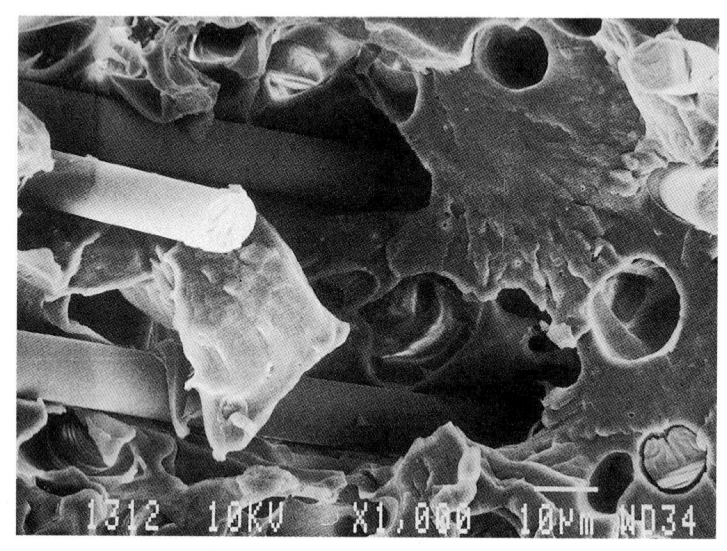

Figure 7.33   Ductile voids in brittle fracture surface. Shell layer. Cryogenically fractured, fatigued specimen, conditioned AL type. Fatigued at 0.7UTS, for 354 cycles, 90% of predicted lifetime.

Chapter 7, Fractography

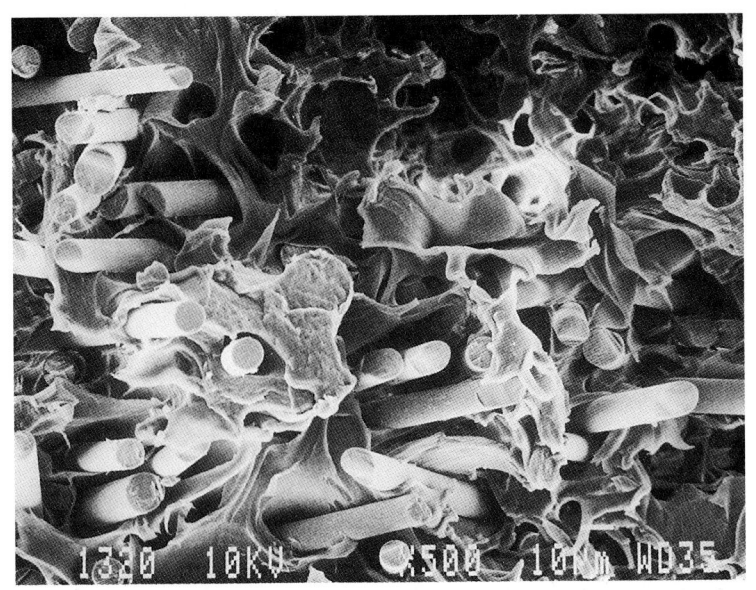

Figure 7.34   *Brittle bridge in ductile fracture surface. Shell layer. Cryogenically fractured, fatigued specimen, conditioned AL type. Fatigued at 0.7UTS, for 354 cycles, 90% of predicted lifetime.*

To be able to determine the site where damage initiates, and where voids are formed, both fracture surfaces of one specimen were compared, Figs. 7.35 and 7.36. Through a tedious procedure the actual voids on one surface could be traced on the other fracture surface. If on one fracture surface a void around a fibre was visible, usually the corresponding void on the other fracture surface was empty: The void was present *at the fibre end*. In less than 10% of the cases, *both corresponding voids* contained a fibre end, indicating that this was a void that was formed at the middle of the fibre. Question remains whether this fibre was broken in cryogenic fracture or during fatigue. This can not be answered using the current method.

In Fig. 7.34 a location is seen where the damage has advanced, a small brittle part in a microductile area can be seen. This brittle part was broken cryogenically. Brittle parts like this do not occur in regular fractographs, and must therefore have been bridges that were still connecting the two crack surfaces. Theoretically it is possible that the crack was bridged also by single fibres, but fracture of these bridging fibres in fatigue can not be distinguished from cryogenic fracture, because the fracture of glassfibres is always brittle. The small area broken in cryogenic fracture, (in this case containing a fibre) bridges the crack. Which makes it possible that the crack can still take up some load. The explanation why this part deforms more easily as compared to the rest of the damaged zone, probably is existing damage on levels above or below the level visible in the cryogenic fracture surface.

Figure 7.35 Ductile voids in brittle surface. Shell layer. Cryogenically fractured specimen, conditioned AL type. 0.7UTS, 354 cycles = 90% of predicted lifetime.

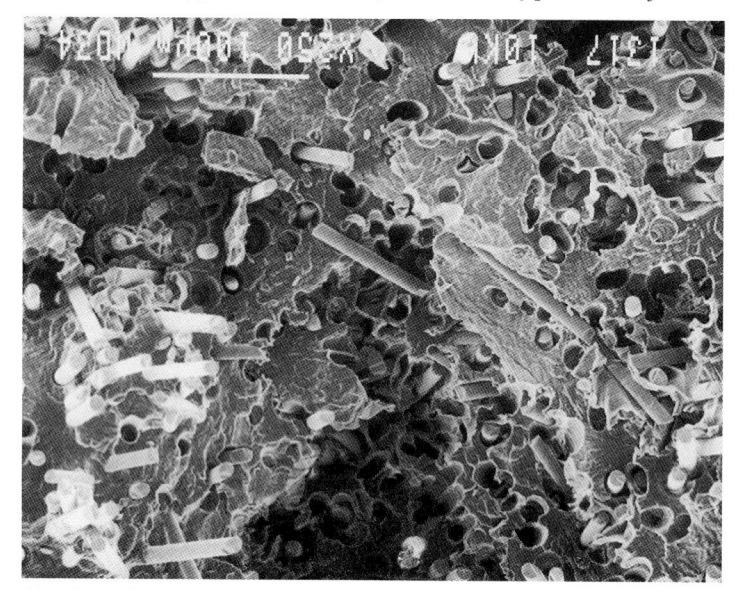

Figure 7.36 Matching fracture surface of Fig. 7.35, 250x.

Figs. 7.33 and 7.34 of course only give details of the damaged zone, after cryogenic fracture. We are interested in the *distribution of the damage* over the specimen. In Fig. 7.37 a map of the corner of one cryogenically broken, fatigued specimen is shown. This is the *only part* of the

Chapter 7, Fractography   121

cryogenic fracture surface of this specimen where ductile behaviour could be observed. Close to the corner the material is entirely microductile (as in Fig. 7.2), except for some small zones that show microbrittle behaviour. These areas are bridges between the two crack surfaces, and obviously were present especially at the surface of the specimen, where the material can deform more easily. This is due to the lower constraint and the more random fibre orientation in the skin layer. Outside this ductile (cracked) zone a transition zone to the outer brittle zone is present. In this area voids around the (debonded) fibres are seen, as in Fig. 7.33.

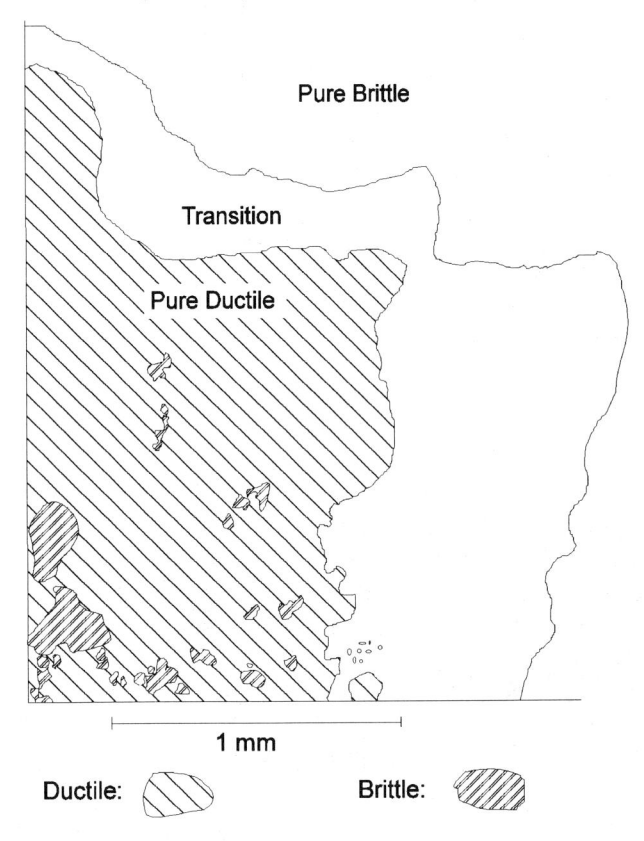

Figure 7.37   Map of corner showing microductility, of cryogenically fractured, fatigued specimen, conditioned AL type. Fatigued at 0.7UTS, for 354 cycles, 90% of predicted lifetime.

A final important observation is that not in all transverse cryogenically broken, fatigued specimens damage could be observed. Approximately 50% of the cryogenically broken specimens showed microbrittle behaviour over all of the fracture surface. No ductility could be observed in these specimens, although after fatigue, stress whitening lines were visible on the specimen. The cryogenic fracture does not tend to follow the path of existing fatigue damage. Also for the specimens where ductility was observed on the cryogenically broken fracture surface, this was observed only on a small part of the fracture surface. The damage that is the consequence of the fatigue loading, therefore is *confined to certain areas*, and is not present throughout the specimen.

7.6.2 Longitudinal cryogenic fracture

For the investigation of the distribution of these damaged zones over the length of the specimen, specimens have to be cut in the longitudinal direction. At first specimens were cut mechanically, either using the microtome technique, or specimens were saw cut, and consequently polished. Both methods imply a mechanical contact with the material, causing damage. This damage can not always be distinguished from damage caused by the fatigue process. Only in extreme cases, where voids were deep enough to exclude that these were formed by mechanical contact in the cutting or polishing process, fatigue damage could be made visible. However, our main interest is the onset of damage, not yet fully developed voids or cracks.

A method without mechanical operations of the surface under investigation was found in confocal laser scanning microscopy. Using this method, it is possible to obtain a view of a plane *under* the surface of the specimen. However, the contrast was not sufficient to distinguish voids or damage, other than close to the specimen surface. This damage was induced by the polishing process. The fibres can be distinguished from the matrix up to a depth of approximately 80 μm.

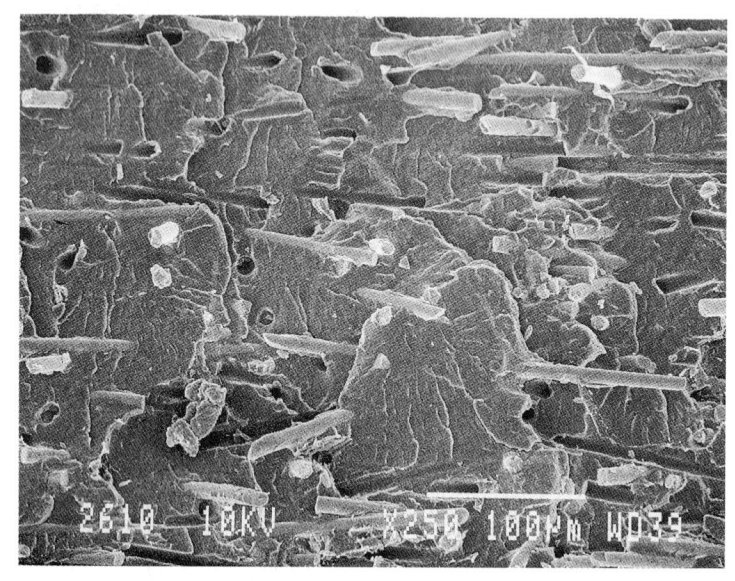

Figure 7.38   Longitudinal cryogenic fracture, fatigued specimen. No microductility can be observed. 250x.

The method that finally did give results, was by using cryogenic fracture to uncover the fatigue damage inside the specimen. From specimens that were fatigued (and had fractured) the remaining part of the specimen, not broken in fatigue, was cut from the shoulders of the specimen. In this remaining part of the narrow section of the specimen, a saw cut was made in longitudinal direction, up to half the thickness of the specimen. Consequently this specimen was

Chapter 7, Fractography 123

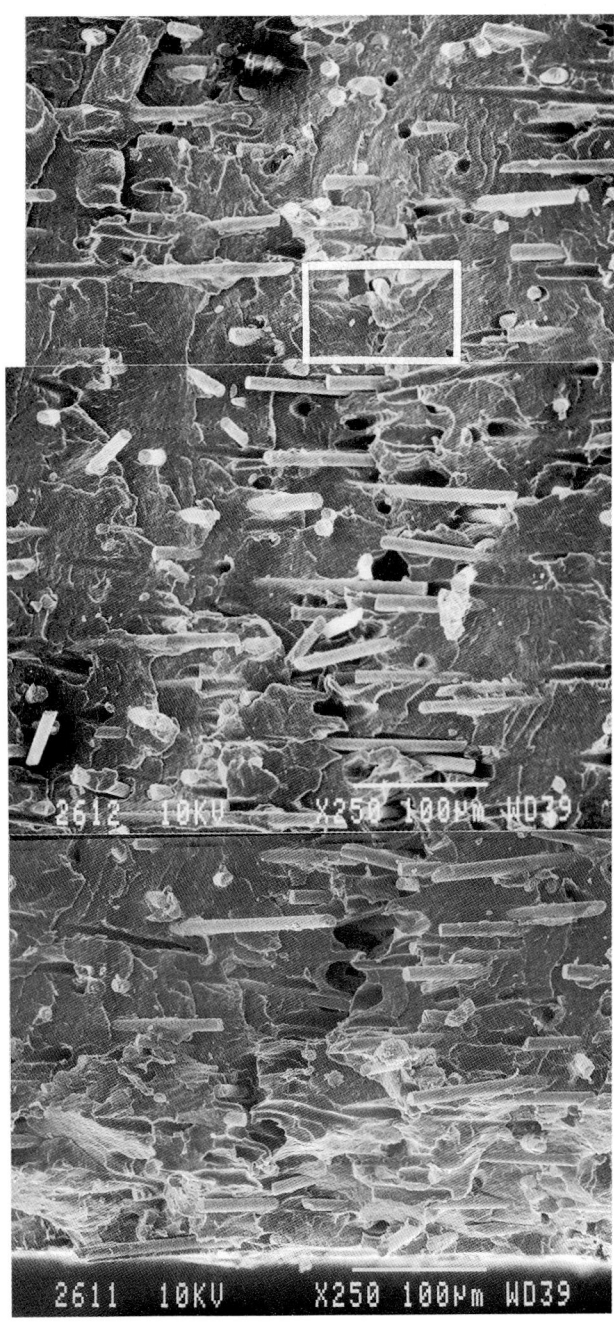

Figure 7.39   *Side view of a fatigue "crack" in a cryogenic fracture surface. Fatigue load from left to right. BL type, 5.75mm thick, 0.55UTS, 3633 cycles to fracture. Detail of the white rectangle is shown in Fig. 7.40.*

cooled in liquid nitrogen, and broken in 3-point bending. The resulting fracture surface is parallel to the loading direction in fatigue, and perpendicular to the surface of the fatigued specimen.

The first observation that can be made from the fractographs, is that most of the cryogenic fracture does not show any sign of matrix ductility at all, as in Fig. 7.38. In 95% of the surface no voids can be seen. This corresponds with the results found in the transverse fracture.

Secondly, areas where a fracture has developed can be found, as in Fig. 7.39. Here a crack can be seen, which is open below, while on the upper part of the photograph only some slight damage can be observed. In Fig. 7.40 a detail is shown of a fibre, transverse to the fatigue load, which is debonded and a void has formed around this fibre.

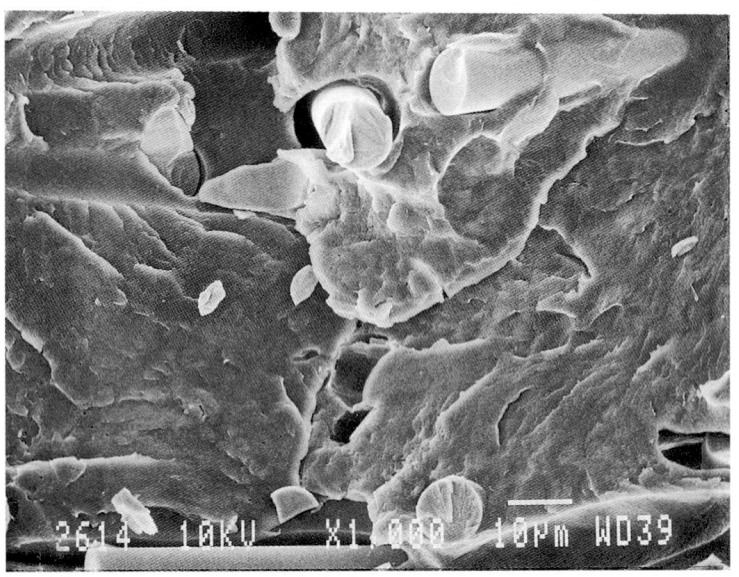

Figure 7.40   Detail of Fig. 7.39. Void around fibre perpendicular to fatigue load direction (from left to right). BL type, 5.75mm thick, 0.55UTS, 3633 cycles to fracture.

In the same specimen only on a limited number of locations deformation is visible. Generally the deformation and void formation is at fibres perpendicular to the fatigue load. However also voids at fibre ends are visible, as is shown in Fig. 7.41. As is visible on the overview photograph of the area around this void, only slight deformations at other, surrounding fibres is present.

Chapter 7, Fractography 125

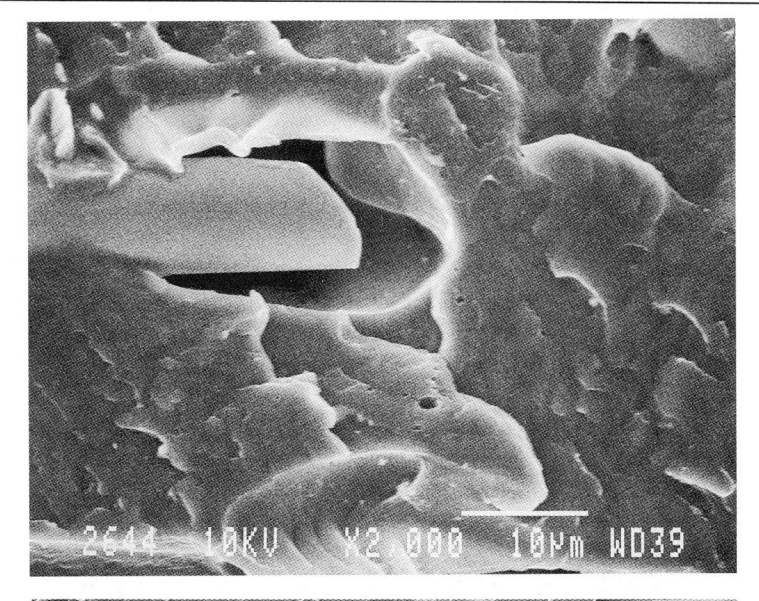

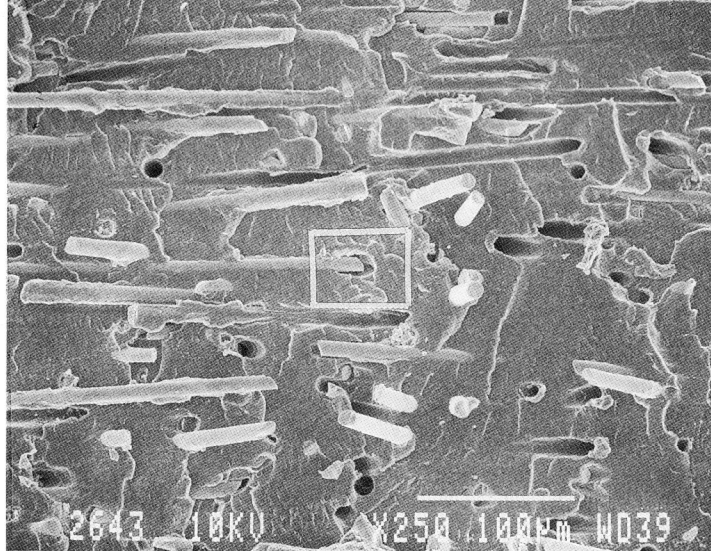

Figure 7.41   *Void around fibre end. On the overview picture (below), not much microductility can be observed. EL type, 5.75mm thick, 0.7UTS, 294 cycles to fracture. Shell layer.*

One detail was observed in a fatigued BL type specimen that is worth noting. In Fig. 7.42 a fractured fibre can be observed. This fibre is probably broken in tension during the fatigue loading. In the photograph some deformation at the other fibre ends is visible, and on the overview photograph it can be seen that the broken fibre itself is debonded at the end. So plastic deformation has taken place, which might have caused fracture of the fibre.

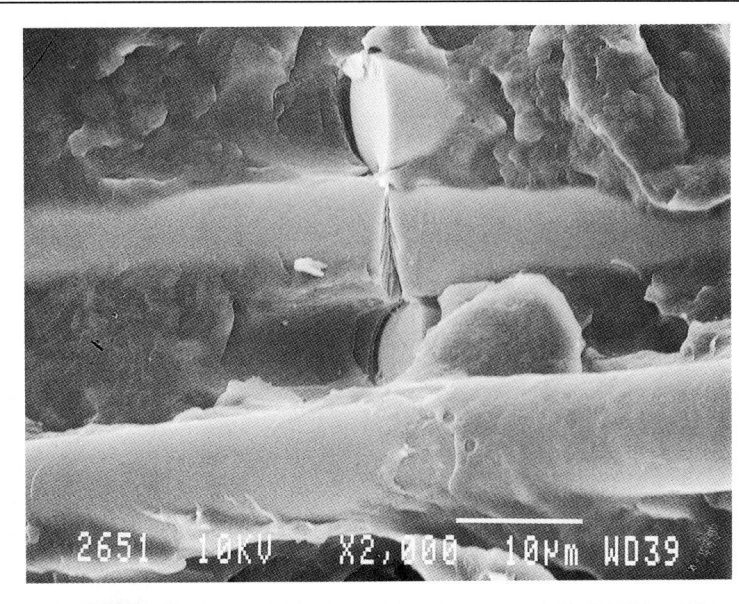

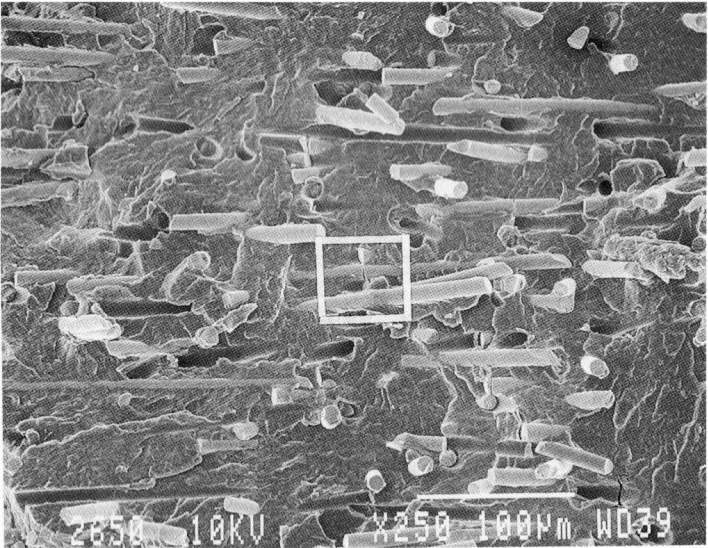

Figure 7.42   *Broken fibre in fatigue load direction. Two fibre ends are visible besides the broken fibre. Hardly any deformation is visible on the overview photograph on the right. BL type, 5.75mm thick, 0.55UTS, 3965 cycles to fracture.*

An estimate of the total deformation in fatigue can be made from the plastic component, as visible in Fig. 7.42, and the elastic component which can be estimated from the elastic Modulus which was measured during the experiment. The visible plastic deformation is 3-5 µm, over a fibre length of 240 µm this is a strain of 1-2%. The elastic deformation, at maximum fatigue load of 3974 N is estimated as 2.3%, using a cyclic Modulus of 3000 MPa. The sum of the plastic and the elastic deformation is between 3.3 and 4.3%. The fracture strain for the glassfibre is 3.5 - 4%, approximately the same value. The proximity of other fibres, as visible in Fig. 7.42, may have induced a bending moment on the fibre as well, increasing the probability of fibre fracture in tension.

7.7   Conclusions

Results for fractography of fatigued specimens agree with fractography reported in literature: In the case of crack growth experiments the fatigue crack growth area shows microductile behaviour, and the degree of ductility increases with crack growth and with increasing stress intensity.
For *tensile experiments* a largely microbrittle fracture surface was found, except for a small part which shows microductility. This also is in agreement with what was reported in literature.
Differences between fracture surfaces from *fatigued* and from *tensile tested* experiments are the higher matrix ductility in fatigue, more fibre breakage in fatigue, due to buckling during the unloading stage of the fatigue load cycle, and a detrimental effect of fatigue on fibre matrix bonding, showing in naked fibres in fatigue, with no matrix material adhering to them.

Different conditioning of the specimens affects two properties of the composite, which lead to a striking difference in appearance of the fracture surfaces. Absorbed water has a plasticising effect on the matrix, showing in an increase of ductility of the matrix with increasing water content. The water also has a weakening effect on the fibre - matrix interfacial strength. This is seen in less matrix material adhering to the fibres, with increasing water content. As the bonding to the fibre puts a constraint on the matrix, making it behave less ductile, the decreasing interfacial strength also leads to an increasing matrix ductility, as this constraint is relieved by debonding.

Concluding mainly from the cryogenically fractured, fatigued specimens, the fatigue damage in the material is confined to areas, and is not distributed evenly over the specimen. Damaged areas behave like crazes, growing perpendicular to the fatigue load direction. Damage begins with debonding and matrix deformation at fibre ends, and at fibres perpendicular to the loading direction. Damage initiates and develops first in the layers where the main fibre orientation is parallel to the loading direction. This can be different from what is reported in literature for material with a higher fibre fraction. In those cases the local fibre fraction in the core, can cause a very low fracture strain in the case the material is loaded parallel to MFD. Fibre fracture in tension does occur, but is a relatively rare phenomenon. Estimates from the ductile - brittle transition areas of specimens are that only 2% of the fibres break in tension.

The damage grows into cracks, which remain bridged if the ductility of the material is sufficient, in conditioned and in 100% wet specimens only. The dry as moulded material shows little matrix ductility, and cracks are not bridged. The less matrix ductility leads to a higher susceptibility of the fibres to buckling. That the dry as moulded material cracks, could also be shown from the size of the microductile part of the fracture surface. Also the increase in matrix ductility with crack growth, similar to what was seen for crack growth experiments, leads to this same conclusion for dry as moulded material.

8. Failure Mechanism

8.1 Introduction

In the previous chapters theory and experimental results were presented, about the failure behaviour of glassfibre reinforced PolyAmide 6. Results of especially the Master Curves for the fatigue behaviour can be used in the engineering of GFPA products,. However, from a scientific point of view, the explanation for the occurrences in fatigue failure are of great interest. Also the existence of the Master Curves for fatigue strength of specimens with different orientation distribution needs to be explained. Depending on the explanation for the Master Curves a prediction can be made of the possibility to apply this method for glassfibre reinforced materials with a different matrix polymer as well.

Failure mechanisms for tensile fracture have been *discussed in literature* by various researchers [Bohse,92, Sato,84, Sato,82], as is the case for fatigue crack propagation [Karbhari,89, Lang,87b, Hertzberg,80, Piggott,80, Dally,69].

In *tensile experiments* the failure mechanism is generally assumed to consist of the following steps [Sato,82], as visualized (for a model system) in Figure 8.1.
1   Cracks initiate at fibre ends
2   Cracks propagate <u>in the matrix</u> along the interface, thus leaving a thin layer of matrix material adhered to the fibre. Only in systems with poor fibre-matrix bonding the interface itself fails.
3   Matrix cracks grow from the interfacial cracks, possibly after generation of matrix plastic deformation [Sato,88].

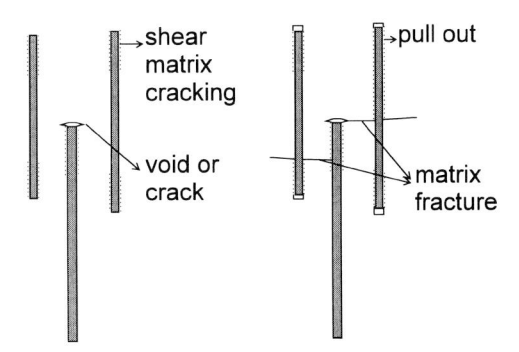

Figure 8.1   Failure mechanism in a tensile experiment. Left: Initiation of crack. Right: growth of crack.

In *fatigue* four stages are usually distinguished [Lang,87b]:
1   Initiation of local weakenings due to cyclic deformation, generally initiating at the locations of highest stress intensity, the fibre ends [Piggott,80].
2   Initiation of (micro)crack.

3   Crack growth due to cyclic loading. Local modes of crack extension depend on local fibre orientation, matrix ductility and the degree of interfacial adhesion [Lang,87b]. The mechanisms during breakdown of the composite are: Fibre matrix separation along the interfaces of fibres oriented parallel to the crack, deformation and fracture of the matrix between fibres, fibre pull-out, and fracture of transverse (to the crack direction) fibres [Malzahn,84].
4   Fast (instable) crack growth in the last load cycle, which should be comparable to failure in a tensile test.

This approach does not account for the orientation distribution inside the material. Hertzberg reports that damage initiates first with debonding of fibres perpendicular to the load direction [Hertzberg,80]. Damage consequently grows into regions with fibre orientations at a smaller angle to the load.

Dally reports the mechanism mentioned above for SFRTP PA6 (dry as moulded) [Dally,69]. In fatigue the damage was seen to be extremely local, abundant and uniformly distributed. The crack propagates by debonding, with limited tendency for crack propagation in the matrix. Many localized regions of cracking extended and coalesced to form a larger cracked area. In a system with *PE matrix*, which is more ductile and hardly bonds to the fibres, he reports an entirely different mechanism. Massive debonding reduces the glass fibres from reinforcement to unbonded inclusions, giving rise to a sharp drop in Modulus. The larger strains are accommodated by the matrix without failure. This redistribution of strain causes the debonded region to progressively enlarge. In measurements where tensile strength of fatigued specimens (but not fatigued until failure) was determined, no decrease in strength could be found, in contrast to the PA6 system. This process of general degradation rather than a dominant crack was also reported by Mandell [Mandell,83] for unnotched specimens.

8.2   Explanation of failure mechanism

As is noted above, and as was seen also in the fractography results presented in Chapter 7, the failure mechanism in fatigue depends on the conditioning of the GFPA. Therefore conditioned, 100% wet (saturated) and dry as moulded materials will be discussed separately. As material with improved fibre - matrix bonding was investigated, we will take a separate look at the failure mechanism for this case.

8.2.1   Conditioned material

Most of our research has been done on conditioned material, assuming that this will be the condition in which it will be used mostly. In Chapter 7 the fractographic evidence for the failure mechanism that is presented here (Fig. 8.2) was discussed. It is based on and explains the experimental results. The failure mechanism consists of the following stages:
1   Initiation of damage at the fibre ends.
2   Growth of this damage into voids, accompanied by debonding.
3   The voids grow into microcracks, which may remain bridged by either drawn matrix material or unbroken fibres.

Chapter 8, Failure Mechanism 131

4   The debonding relieves the constraint to which the matrix was subjected, which can therefore deform much more easily, forming bridges between the crack walls.
5   The bridged crack grows, until a critical size is reached, and the specimen fails.

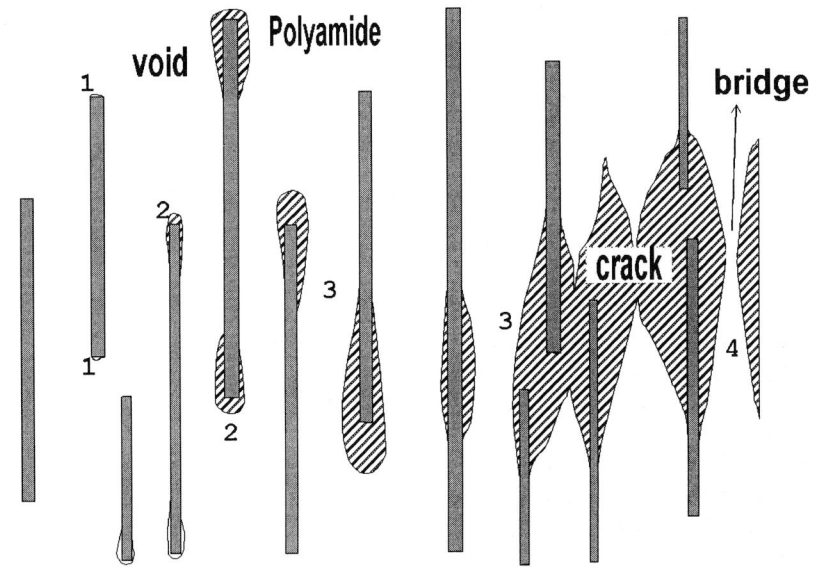

Figure 8.2   Failure mechanism in fatigue, development of damage from left to right. The numbers are explained in the text.

In this case damage of the layers with fibre orientation predominantly in loading direction dominates the fatigue failure. Layers where the fibre orientation is perpendicular to the loading direction show a high degree of fibre - matrix debonding and plastic deformation of the matrix in between fibres.

A similar craze-like behaviour in fatigue was reported by Dibenedetto and Salee [Dibenedetto,79], notably for fatigue of compact tension specimens of graphite fibre reinforced PA 66. These specimens did not show any crack-growth until the last few load cycles before failure.

The existence of the bridged cracks can be concluded from the following observations: The stress whitening lines visible at the surface, and after microtoming also inside the specimen, indicate plastically deformed matrix material. The matrix material can not deform to such an extent, without debonding from the fibres. This because the glassfibres can deform only 3.5 - 4% maximum. On the other hand the plastic deformation causes a tensile stress on the fibre - matrix interface (See § 8.3), accelerating the debonding process. The plastic deformation is also visible in fractography, where a high degree of matrix drawing can be seen, and fibres totally debonded from the matrix. These plastically deformed zones however are no cracks. This can be concluded from the observation that during the strength profile measurements, presented in Chapter 6, no foils with strength zero could be found. The strength of foils containing a plastically deformed zone (visible in stress whitening lines) is remarkably lower compared to foils not containing this.

When observing an experiment a growing number of white lines is seen, although none is a real crack. A crack can sometimes be observed in the last but one load cycle, when fast fracture seems to have begun, and is stopped by the decreasing load. Final fracture then takes place in the next load cycle.

Fractography of cryogenically fractured, fatigued specimens (§ 7.6) further confirms the correctness of the model here presented. Voids around fibre ends are abundant. The voids at fibre ends could be actually seen, and are similar to those depicted in Fig. 8.2. Fibres perpendicular to the loading direction are considered to be a weak spot as well. For these fibres debonding from the matrix has been observed repeatedly, as in Fig. 7.40.

A phenomenon that needs to be explained is the low fatigue resistance of the *BL-type* specimens, as shown in Fig. 4.11. These specimens do not fracture perpendicular to the specimen axis, but at an angle of approximately 45°. The local fibre orientation in the core layer corresponds with this angle. A shear mechanism along the main fibre orientation direction is present here. A possible reason why this mechanism gives this low fatigue resistance, compared with the mechanism in case the fibres are perpendicular to the loading direction or parallel to loading direction, may be as follows: Compared with the perpendicular fibres the stress intensity at the fibre ends of these fibres at an angle is higher, and far more comparable with the stress intensity in the case of parallel (to the loading direction) fibres. However, the stress for debonding for the fibres at an angle is not shear, as in the case of parallel fibres, but also tension, as for perpendicular fibres. Thus for fibres at an angle occurrence of both a stress intensity at the fibre end, as well as a tensile debonding stress may be the cause for the low fatigue strength of the composite.

Another feature that could be seen in fractography was the occurrence of fibres broken in tension. These fibres are not broken in the unloading stage of the fatigue load cycle, through buckling or bending which are often reported, but in tension: see §7.3.3.2 and Fig. 7.42. This is a relatively scarce phenomenon though: approximately 2% of the fibres are broken in tension.

The Creep speed V_c (§ 4.3.3) is the reflection of the initiation, growth and opening of all the bridged cracks over the length of the specimen. A low V_c indicates a low growth rate of the bridged cracks, resulting in a higher number of cycles until the critical size of one fatal bridged crack is reached and the specimen fails. This explains the correlation at one load level. That the relation between V_c and N exists for all load levels, indicates that the proposed mechanism is valid for the entire load range tested. That this is the case can be understood by considering the effect of the fibres on the stress inside the material. Fibre misalignment, fibres touching and fibre ends will highly affect the local stress, increasing the local stress to the minimum stress level required for initiation of damage. This level will be reached locally at all load levels, but at a decreasing number of locations with decreasing load. Therefore initiation of bridged cracks and growth of these will occur at all load levels used in this research.

8.2.2 Saturated material

The mechanism in fatigue for the 100% wet material is similar to that in conditioned material, which can be concluded mainly from the fractographic results. The deformation of the polymer matrix is even higher than it was for conditioned material, giving more possibility for bridges to form. The failure mechanism for tensile experiments does change. This can be concluded from the very high percentage ductile area on the fracture surface (Figs. 7.19 - 7.20) of a tensile experiment. It seems that the interfacial strength has been lowered by the absorbed water, to a value below the minimum shear strength at which the interface itself fails, not the matrix. This can also be concluded from the observation that, if the material is 100% wet, the experimental material has a higher tensile strength than the standard grade. No further investigations have been done on this subject, but it seems that the fracture mechanism in a tensile experiment of wet material is similar to the mechanism in fatigue for a conditioned specimen.

8.2.3 Dry as moulded material

In fractography the sharp difference in matrix ductility between dry as moulded and conditioned material was observed, Compare Fig. 7.16 with Fig. 7.11. The comparison has been visualised in Fig. 7.18. In the dry as moulded case hardly any ductility is observed, the surface is more crack-like than in the case of the conditioned material. As this ductility is needed for the formation of bridges between the crack walls, the dry as moulded material will form cracks that are *not bridged*. Thus the mechanism for dry as moulded material is in accordance with the mechanism as reported in literature, and mentioned in the introduction of this chapter. This can also be concluded from the measurements of the size of the microductile area of the fracture surface presented in §7.4.2, where the dry as moulded material was seen to be notch sensitive, while this is not the case for the conditioned material.

8.2.4 Material with improved bonding

The fatigue mechanism for material with improved bonding is not different from that for the standard material, neither for dry as moulded nor for conditioned material. Matrix - fibre bonding is better, which leads to a slower development of debonding in fatigue. However, the size of the microductile zone on the fracture surface is the same for both cases [Horst,96b], showing that no elementary changes in the failure mechanism occurs. The same type of damage occurs, but the damage takes more load cycles to develop.

8.3 Modelling of tensile debonding

In earlier published theories concerning the failure mechanism, the fibre - matrix interface is considered to fail in a shear mode, and the matrix itself cracks. Contrary to this, the theory here presented and based on experimental observations, considers the tensile (fatigue) strength of the interface of great importance, not just the pull-out shear strength.
Imagine a system of perfectly aligned fibres, loaded parallel to the fibre orientation direction, Fig. 8.4. At the fibre ends the local stress will be high. Damage will occur at the fibre end, followed by debonding in a shear mode at the ends of the fibres, where shear stress at the interface is highest. The unloading of the fibres results in an increasing load on the matrix, which will deform. Deformation of the matrix is inhibited by the bonding to the fibres, which makes lateral contraction of the matrix, needed for deformation, impossible. This results in a tensile stress on the fibre - matrix interface making it fail in a tensile (fatigue) way.

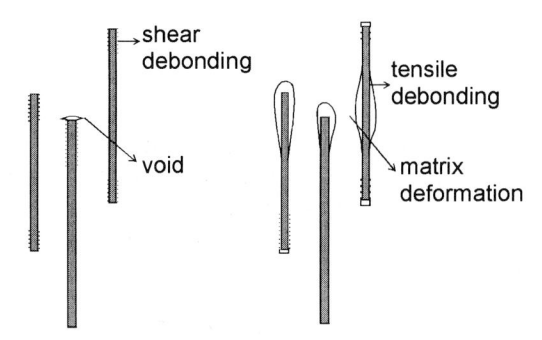

Figure 8.3 Tensile debonding of the interface, due to the lateral contraction if the matrix is deformed. Loading direction is vertical, parallel to the fibres.

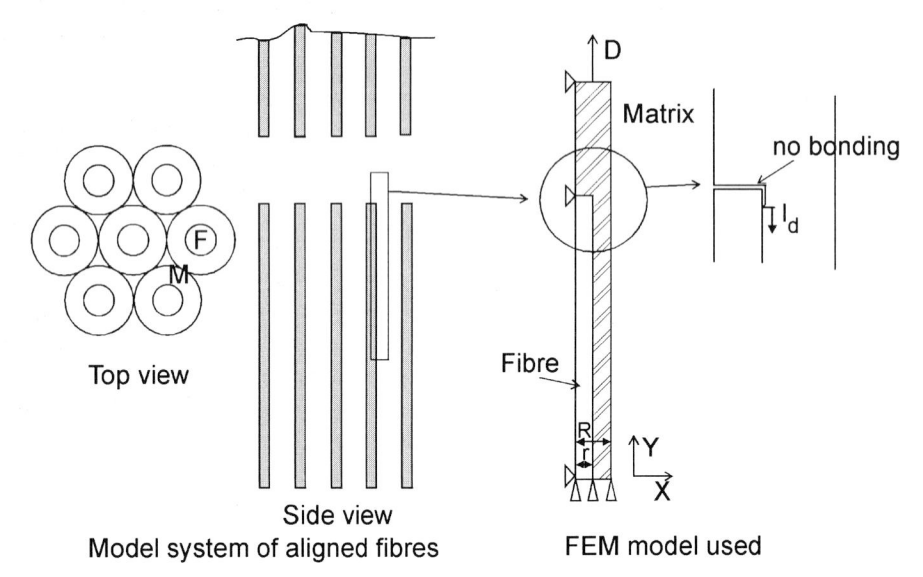

Figure 8.4 Model used for FEM calculations. A quarter of the fibre is modelled in an axisymmetric model. Bonding is assumed to be absent at the fibre end in one calculation, and also along part of the fibre in another, see magnification of fibre end. Loading direction parallel to fibres.

The existence of the tensile stress on the interface could be proven in a preliminary study using Finite Element Method modelling. An axisymmetric model of half a fibre was made, thus reducing the finite element model to a 2D problem. No bonding between fibre end and matrix was assumed, while in the main model discussed here also a small part along the fibre, near the fibre end was debonded. See detail in Fig. 8.4. Thus the debonding stresses are studied for a situation where some debonding along the fibre is already present. A regular hexagonal distribution of the fibres was assumed. See Fig. 8.4. All fibre ends lie in one plane, to be able to simulate the high deformations at the fibre ends, as seen in Fig. 7.41. The dimensions are:
r   = 5μm

Chapter 8, Failure Mechanism

R = 10 μm
0.5l = 150 μm (length of fibre)
0.5l_m = 200 μm (length of model)
v_f = 0.19 (fibre volume fraction)

An appropriate mesh was used, with a finer mesh in the region of the fibre end (In Fig. 8.6 the fine mesh at the fibre end is shown). Thus a total of approximately 2000 elements was needed in these preliminary calculations. The calculation was done using the Ansys program, release 5.3 for Windows NT. The element type was Plane 82, axisymmetric, 8 nodes. The whole model is quite similar to the one used by Brockmüller [Brockmüller,95].

Crucial in FEM modelling are the boundary conditions. In the FEM model in Figure 8.4 the boundary conditions are:
- On the left side (X=0): axisymmetry
- At the bottom (Y=0): normal symmetry applies: displacement in Y-direction is 0
- At the top the (Y=l_m): displacement is uniform and in Y-direction (and the input for the calculation).
- At the right side (X=R) the boundary condition needs some consideration. If we leave the right side free, we consider there is no connection between the matrix surrounding two different fibres, see the top view. Lateral contraction of the matrix would leave a void *inside the matrix*, between the fibres. As we know from fractography, no voids in the matrix occur. Thus the lateral contraction of the right side of our model must be inhibited. However, assuming the displacement of the right side (in X-direction) to be zero would deny the lateral contraction altogether. Therefore it is assumed that the right side remains *straight*, the displacement in X-direction is uniform over the entire length. The value of the displacement is dictated by the lateral contraction far from the fibre end. This because we assume the specimen to contain only one voided area, the rest of the specimen is free of voids, and the lateral contraction of this is therefore the low value present at the middle of the fibre (at the bottom of our model). The stress in X-direction must be zero at this location.

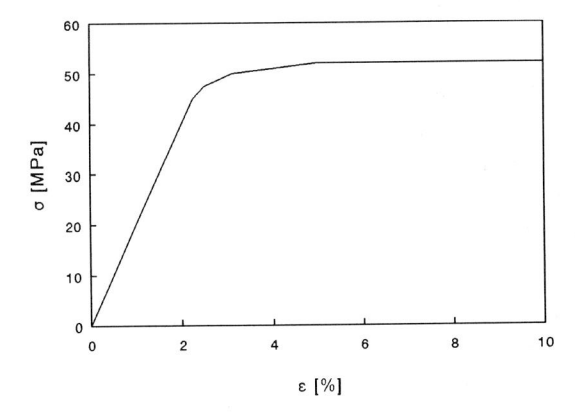

Figure 8.5   Stress - strain curve for the matrix.

As plasticity of the matrix was observed, it is essential to include this in the material properties for the matrix. In Fig. 8.5 the stress-strain curve for the PolyAmide matrix, as used in this model is shown. Initial Modulus is 2000MPa and poisson's ratio $\nu = 0.42$. For the fibre the Modulus is 70 Gpa, and $\nu = 0.35$.

Calculations were executed using a stepwise increase of the displacement. Until a maximum displacement of 4 ($\equiv$ 2% strain) no computational difficulties were encountered. In Figure 8.6 the development of deformation at the fibre end is shown. Compare with Fig. 7.41 of the actually photographed void at the fibre end, on the fracture surface of a cryogenically broken specimen. In all three graphs the upper 4 elements at the fibre end are *debonded* from the matrix. For higher displacements the local deformations near the fibre end lead to problems with the calculation. The results for both models (with and without debonding along the top of the fibre) are very similar. For low imposed strains (< 0.5%) the matrix stresses are largely in the elastic region. Therefore the effect of lateral contraction is fairly small, and the shear stress along the

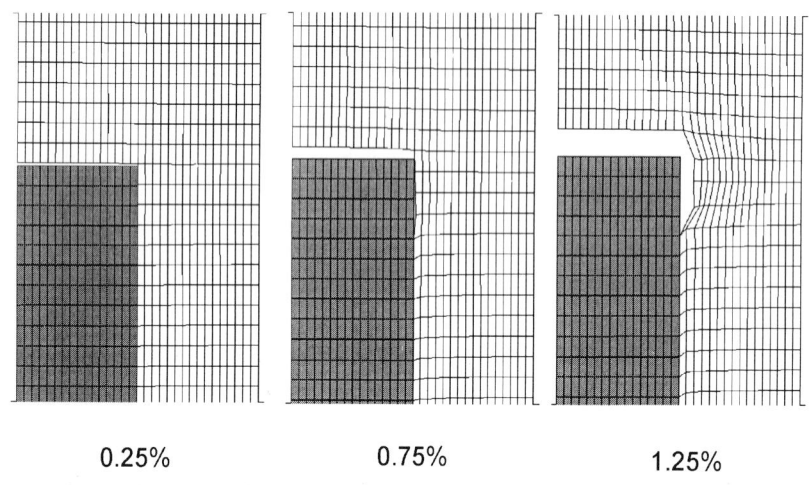

0.25%   0.75%   1.25%

Figure 8.6 *Deformations at the fibre end. Shown is the deformed mesh for three values of strain. These are the overall strains imposed on the model.*

fibre is higher than the tensile stress σ_x on the fibre, see Fig.8.7a. With increasing displacement, and therefore increasing plastic deformation of the matrix near the fibre end, the tensile stress in x direction (perpendicular to the fibre axis) becomes higher than the shear stress, close to the debonded area of the fibre (Fig. 8.7b and c). The tensile stress shows a high peak near the debonded area, and decreases rapidly with increasing distance from the debonded area, l_d. The shear stress is fairly constant, and decreases much more gradually. Due to the opening of the void at higher strains, the principal stress direction near the debonded region changes in a direction more perpendicular to the fibre axis. Consequence of this is that the highest shear stress does not occur near the debonded area. In Fig. 8.7 the stresses near the fibre end are shown for the case with debonding along the fibre top. For the case where only the fibre end was debonded, results are quite similar, and are therefore not shown.

We can conclude therefore that for the cases where plastic deformation of the matrix is present, as in fatigue failure, not just the pull-out (shear) strength of the interface, but mainly the *tensile strength* of this is of importance. The theory that a tensile stress at the interface may exist has

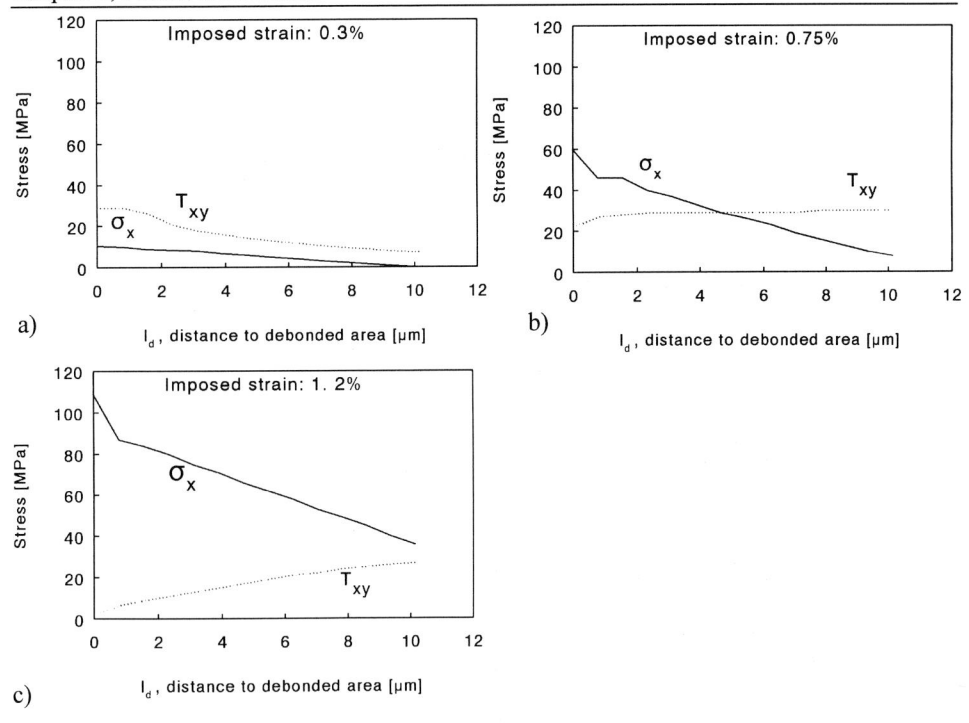

Figure 8.7   *Development of shear (τ_{xy}) and tensile (σ_x) stresses along the fibre with distance to the debonded area l_d (near the fibre end), at three different values for the imposed strain.*

been confirmed. Moreover, the tensile stress can be higher than the shear stress.
In case of purely elastic matrix behaviour, the shear stress at the interface shows a much higher peak near the debonded area, also because the void does not open so much as for the elastoplastic case. The shear stress at the interface for the elastic case is **higher** than the tensile stress [Horst,97b], although this latter is still considerable.

8.4   Implications of the failure mechanism

One of the major implications, especially for materials research, comes from the difference in role of the interface for fatigue and tensile experiments. Is the interfacial strength above a certain level, which for dry as moulded or conditioned GFPA normally will be the case, the interface *does not fail* in a tensile test. The matrix fails, close to the fibre. In fatigue the interface itself fails [Horst,96b]. Thus an *improvement of interfacial strength* does *not* improve the *tensile strength*. Therefore in material development tensile experiments alone are not adequate for evaluation of the strength of the interface, fatigue experiments provide a better means for this.

This leads us to the subject of improvement of the material behaviour. It will be discussed per constituent; Glassfibre, PolyAmide matrix and the Interface between fibre and matrix.

The GlassFibres

The properties of the fibres can not be improved much, assuming that the glassfibres will not be replaced by a different grade of fibres, like carbon fibres. For improvement thus remains the geometry of the fibre, an increase of the fibre aspect ratio will improve both tensile as well as fatigue properties. An increase of the fibre aspect ratio can be accomplished either by decreasing the fibre diameter [Sato,88], or by increasing the fibre length.

However, a practical minimum of the fibre diameter is 7 µm, so the possible improvement of properties is limited. Increase of the fibre length is possible by using the pultrusion technique to produce the compound [Karger-Kocsis,88], combined with adapting the mould design and the injection moulding parameters for minimum fibre fracture due to the injection moulding process. Possible fibre lengths are 1 - 1.5 mm, which compares very favourably with the 0.25 to 0.35 mm of the standard material. A drawback of fibre length increase is *fibre clustering*. The fibres tend to group together in clusters during injection moulding [Karger-Kocsis,88, Ranganathan,90] or compounding, which decreases the effective fibre aspect ratio, as the diameter of the fibre cluster is the determining dimension for the aspect ratio.

The fibre volume fraction can be increased to increase the properties. The standard grade used in this research contains 20%vol. (30%Wt.) of glassfibres, which can be increased to about 35%vol. maximum. However the higher fibre content makes injection moulding more difficult, as is also the case for long fibres, and with increasing fibre content fibre clustering increases as well.

The matrix

Increase of the matrix elastic Modulus and strength has a direct effect on the composite properties. This can be seen in chapter 6, where a decrease of matrix strength at the surface of the specimens, due to a higher amount of absorbed water, is reflected in the composite strength. Increase of matrix crystallinity by changing the injection moulding parameters or by adding a nucleating agent to the PolyAmide will increase the composite properties and decrease the amount of water that can be absorbed. As the matrix crystallinity is low especially at the surface of the moulding, increase of crystallinity in this area will have a considerable effect when the product is *loaded in bending*.

The interface

Improvement of the interface strength will have an effect in fatigue only, and not in tensile or impact loading. This contrary to the improvement of matrix or fibres, which also affects the tensile properties, and often influences the impact resistance negatively. Improvement of the interfacial strength can be accomplished by using different chemical coating (seizing) of the fibre.

A point which may have some effect is to increase the bonding at the fibre ends. As the fibres are broken during compounding and injection moulding, the fibre ends are not coated, and thus have minimum bonding strength. Adding a coupling agent to the matrix will improve this bonding, although it is not known to what extent this will improve the composite properties. It is possible that the stress raise at the sharp corners of the fibre ends is an effect which is so strong that adding bonding to the fibre ends may have no or little effect.

Chapter 8, Failure Mechanism

The last possibility for improvement of the properties is the adding of an interphase, a different polymeric material between fibre and matrix. This is a practice which is already used in continuous fibre reinforced polymers [Venderbosch,93], while for short fibre composites a theoretical study of the influence of an interphase on mechanical properties has been made by Monette et al. [Monette,93]. An interphase of a more ductile material will decrease the stress intensity effect of the fibre ends on the matrix. Also the fibre - fibre distances will increase, decreasing the high stress intensity which will occur if fibres touch or are close together. A more ductile interphase will lead to a lower stiffness of the material, which may be unacceptable.

To put the interphase on the fibres will be the difficult part. Possibility is to blend a different polymer with the PolyAmide, which has to settle preferentially on the fibres during solidifying. Needless to say that the bonding between fibre and interphase and between interphase and matrix must be good.

Thus various routes are still available to improve the tensile and fatigue behaviour of the material. Normally, if the tensile behaviour will be improved, the fatigue behaviour will improve as well. It must be noted however, that some of the possible changes of the material, e.g. increasing matrix properties, can give a decrease in impact strength.

8.5 Modelling of fatigue behaviour using Master Curves

The Master Curves for fatigue, as presented in Chapter 4, need to be explained, Especially to know if the method of normalisation of Wöhler curves will be applicable for fibre reinforced materials with a *different matrix polymer*. The fact that the normalisation procedure works for *differently conditioned GFPA* gives an indication that Master Curves can be made for other grades of short fibre reinforced polymers as well. The differently conditioned GFPA materials have a different failure mechanism (§ 8.2), showing that the normalisation method of making Master Curves is *independent of the failure mechanism*.

The explanation for the Master Curves was found in the strain at fracture in fatigue combined with the stress distribution through the thickness of the specimen. In Fig. 8.8 the strain at fracture is shown for dry as moulded material.

It can be seen that the strain at fracture is independent of the specimen type, when fatigued at the same percentage of tensile strength UTS, Fig. 8.8a. The strain at fracture thus is independent of the fibre orientation inside the specimen. A similar curve was found for the case of 2 mm thick specimens, though the strains at fatigue fracture for the 2 and 5.75 mm thick specimens are not the same.

This independence of the fracture strain in fatigue of the specimen type, can be explained by assuming that a critical layer exists, which fails in fatigue at a certain critical strain ϵ_{fract}. The thickness of this critical layer does not influence the critical strain of the layer, and thus is

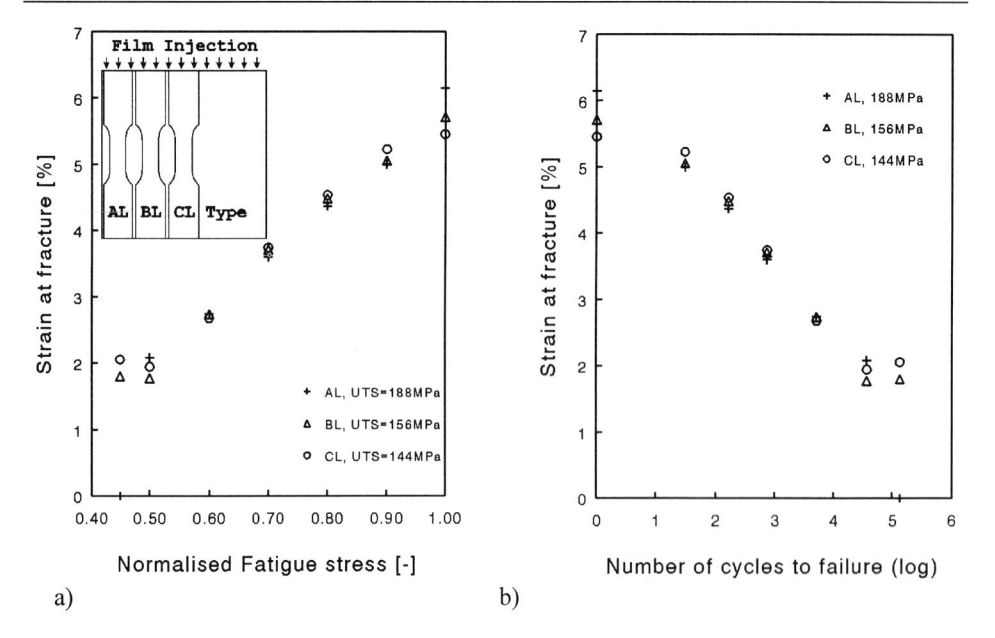

Figure 8.8   Strain at fracture for 3 types of dry as moulded, 5.75 mm thick specimens. a) as function of normalised fatigue stress, b) as function of lifetime.

independent of the specimen type. The nature of the critical layer will be discussed later. The Master Curves can be explained using the following steps:

1)   The critical strain is attained in a certain lifetime (number of cycles), Fig. 8.8b, which is also independent of the specimen type: The thickness of the critical layer is of no influence.

2)   The critical strain is attained during the lifetime due to the application of a certain cyclic strain ϵ_{cycl}. The critical layer shows cyclic creep (ϵ_{creep}) until the critical strain is reached. The thickness of other layers is of little or no influence on the number of cycles in which the critical strain ϵ_{fract} is reached. $\epsilon_{fract} = \epsilon_{cycl} + \epsilon_{creep}$

3)   From the existence of the Master Curves, it is known that the lifetimes of different specimen types are the same, if these are loaded at a fatigue load equal to the same percentage of tensile strength UTS. Thus different specimens at different stresses, but at the same percentage of UTS, show the same critical strain, and the same lifetime. As a certain critical strain ϵ_{fract} is caused by the same cyclic strain ϵ_{cycl}, the conclusion is that a loading with a certain percentage of UTS is in fact the loading with the same cyclic strain ϵ_{cycl}.

4)   Thus the cyclic strain must be the same for different specimen types, when loaded at the same percentage of UTS. This can be explained by the build-up of the tensile strength and the fatigue stress over the different layers in the thickness of the specimen. The

Chapter 8, Failure Mechanism

distribution of the elastic moduli over the different layers is the same for the high and the low stresses / strains.

In equations:

$$UTS = \sum \epsilon_b E_i d_i / \sum d_i = \epsilon_b \sum E_i d_i / \sum d_i \qquad 8.1$$

$$\sigma_{fat} = \sum \epsilon_{cycl} E'_i d_i / \sum d_i = \epsilon_{cycl} \sum E'_i d_i / \sum d_i \qquad 8.2$$

$$\frac{\sigma_{fat}}{UTS} = \frac{\epsilon_{cycl}}{\epsilon_b} \frac{\sum E'_i d_i / \sum d_i}{\sum E_i d_i / \sum d_i} = \frac{\epsilon_{cycl}}{\epsilon_b} \sum \frac{E'_i}{E_i} d_i / \sum d_i \qquad 8.3$$

with:
- UTS = tensile strength
- σ_{fat} = fatigue stress
- ϵ_b = Elastic strain at fracture in a tensile test
- ϵ_{cycl} = $\epsilon_{fract} - \epsilon_{creep}$: cyclic strain in fatigue
- E_i = Elastic (secant) Modulus of layer i
- E'_i = Modulus of elasticity in fatigue
- d_i = thickness of layer i

Note: ϵ_{cycl} and E' are not constant. E' decreases with fatigue (maximum decrease: 10%), and consequently ϵ_{cycl} increases.

If the stress-strain curve of the material would be linear, the elastic moduli of the different layers would be independent of strain, and their relation would be independent of strain. However, as is shown in Fig. 8.9 for conditioned specimens, the stress strain relationship is curved.

What we need though for the cyclic strain to be the same value for different specimen types when loaded in fatigue at the same fraction of UTS, is not that the stress-strain curve is linear, but that the proportionality between the elastic moduli in the different layers does not change with strain: E'_i/E_i is constant. This means that the shape of the tensile curves for the different layers must be the same. That this is the case for this material can be seen from Fig. 8.10. Here the same tensile curves of Fig. 8.9 are plotted, except that the load for the CT type specimens is multiplied by a factor of 2.85. The two curves are congruent: The elastic (secant) Modulus for the AL type specimen is a factor of 2.85 higher than for the CT type specimen. What we can see clearly also is that the plastic strain at fracture for the CT type is higher than for the AL type specimen.

As the AL type specimen mainly consists of layers containing fibres in longitudinal direction, and the CT type specimen mainly contains perpendicular fibres, the similarity of shape as visible for these two specimen types, must also be present for the differently oriented layers.

So the conclusion is that, if σ_{fat}/UTS is a certain value, the cyclic strain is the same value, because the ratio $\epsilon_{cycl}/\epsilon_b$ is the same, independent of the specimen type, because the term $\sum(E'_i/E_i)d_i /\sum d_i$ in Equation 8.3 is constant.

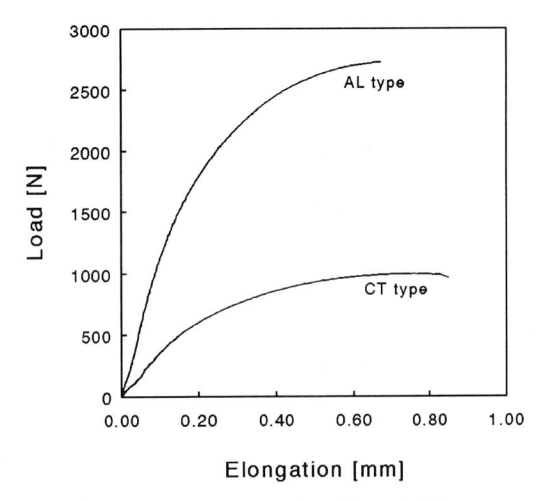

Figure 8.9   Tensile experiment curves for AL and CT type, conditioned specimens.

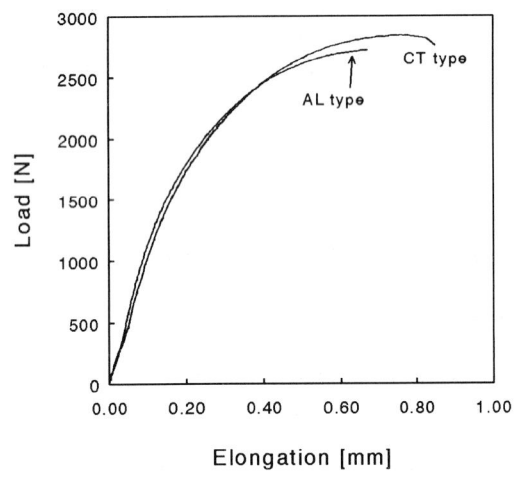

Figure 8.10 Comparison of the shapes of the tensile curves for AL and CT type specimens. The load for the CT type specimen has been multiplied by a factor of 2.85.

About the nature of the critical layer; the assumption is that this is the layer with the fibres aligned with loading direction, because this layer will have the smallest strain at fracture. However, in some articles the core layer, with fibres perpendicular to the loading direction, was reported to fail first [Hitchen,93a and b]. In the case mentioned however, the fibre fraction was 50%Wt, leading to a low content of matrix between the fibres. Locally in the core layer the fibre packing is so high, that little matrix material between the fibres is present. Consequence of this is that the fracture strain locally decreases and fracture of the core occurs. However, the shell layers continue to bear the load, until fatigue fracture of the specimen. Thus the weakest layer, the core layer, which fails first, is not the critical layer in this case, as the final fracture of the specimen is dominated by the *shell layers*.

8.6 Conclusions

The fatigue failure mechanism as presented in this chapter has been concluded from various results and observations. The mechanism changes when the specimens are conditioned differently. For the *dry as moulded* specimens the fatigue failure mechanism is in accordance with that reported in literature, and is a mechanism of:

1   Crack initiation and shear debonding at fibre ends.

2   The crack grows through the matrix, accompanied with fibre fracture, mainly due to buckling in the unloading stage of the fatigue load cycle.

3   If the residual load carrying capacity of the cracked specimen has decreased to the maximum fatigue load, due to the existence of the fatigue crack, the specimen fails in instable crack propagation.

However, for conditioned specimens the situation is quite different, the matrix does not crack, but deforms plastically. Due to this deformation the formation of bridges, connecting the "crack" walls is possible. Less fibre buckling occurs, because the deformation of the matrix enables the fibres to move more freely, as compared to the dry as moulded case. Thus in *conditioned*, as well as in *100% wet material* the fatigue failure mechanism is:

1   Initiation of damage at the fibre ends.

2   Growth of voids at the fibre ends, debonding of the matrix from the fibre occurs.

3   Voids grow into microcracks.

4   The debonding relieves the constraint from the matrix, enabling the matrix to deform. This deformation makes the formation of bridges between the crack walls possible.

5   This bridged crack or damaged area grows perpendicular to the main stress direction, until the residual load carrying ability of the specimen is lowered to the maximum fatigue load, by the presence of the damage. Than the specimen cracks as in a tensile test.

An important consequence of the matrix deformation is the debonding of the matrix from the fibres in a tensile way, perpendicular to the fibre axis, for fibres with orientation in the loading direction. The (plastic) deformation of the matrix is inhibited by the bonding to the fibres. The lateral contraction of the matrix, needed for (plastic) deformation results in a tensile stress on the interface, as shown in § 8.3. This tensile stress can reach values that are significantly higher than the shear stress at the interface.

A further important observation is the difference in interfacial failure site between tensile and fatigue experiments. In fatigue experiments the matrix debonds from the fibre at the interface itself, leaving hardly any matrix material bonded to the fibre. Opposite to this in tensile experiments, the interface itself does not fail, the matrix fails in a shear mode, at a small distance from the fibre, leaving a sheath of matrix material attached to the fibre. Thus in tensile experiments improvement of the fibre matrix interface strength can not be observed, because the interface does not fail.

Methods to improve the fatigue behaviour of the composite system therefore consist of improving the interfacial strength by using different fibre coatings or seizings. Also the introduction of an interphase, a third phase in between matrix and fibres is possible. Of course the fatigue behaviour can also be improved by the known methods to improve the composite tensile strength, like increasing the fibre aspect ratio or volume fraction, or improving the matrix properties by increasing its degree of crystallinity.

The Master Curves presented in Chapter 4, which connect the fatigue behaviour of a specimen with a certain fibre orientation to the tensile behaviour of this specimen can be explained by the critical strain in fatigue. This critical strain at which the specimen fractures is almost independent of the fibre orientation in the specimen, and can be related to a layer in the thickness of the specimen, the critical layer. The existence of the Master Curves can than be explained by the fact that the distribution of the elastic moduli over the layers with different fibre orientation is the same in a tensile and a fatigue experiment.

9. Practical applications of the research

9.1 Introduction

The fatigue experiments on short glassfibre reinforced PolyAmide and the results thereof, as presented in this thesis enables designing for a specified lifetime. Especially the Master Curves introduced to combine the fatigue and tensile behaviour of SFRTP's as depending on the glassfibre orientation can be useful to evaluate the fatigue behaviour of real structures. Preferably *before* the mould is actually build, the fatigue behaviour of injection moulded parts under cyclic loading can be obtained using the following steps:

1. Prediction of the glassfibre orientation throughout the part. Commercial programs already give a prediction of the fibre orientation, although the result is not accurate enough in some cases, see Chapter 2. Advances in this field are expected however in the near future.
2. Using the fibre orientation from step 1, the Elastic Modulus and strength distribution over the part must be evaluated. In Chapter 3 a short introduction on the theory is given. Predictions especially of the elastic Modulus are quite accurate if the fibre orientation is known, and predictions of strength are improving as well.
3. Using FEM the stress distribution in the part must be calculated, resulting from the anisotropic and inhomogeneous elastic moduli and the loading characteristics.
4. Combining the stresses from step 3, and the strengths from step 2 gives the fraction of UTS that is reached locally. This can be compared with the Master Curve to evaluate if the required lifetime can be attained.

These steps of course can be only used in the last stages of the design, when all major dimensions are known. In the remaining of this chapter methods and design rules will be provided that can be used in earlier stages of design.

First the results of the research are summarised in so far the fibre orientation has an influence on the fatigue behaviour. Subsequently possible methods to influence this orientation are discussed. This can be done using the design of the mould, and in a lesser extent by varying the parameters of the actual injection moulding process.

9.2 Design rules for fatigue

The general results as to the influence of the fibre orientation on the fatigue behaviour are similar to its influence on strength. Fibres aligned in the load direction give a better fatigue behaviour as compared with fibres perpendicular to the loading direction. A distinction must be made here between loading in *tension* and in *bending*. As the research was focussed on tension only, *loading in tension* will be our main concern here as well. The effects in bending are mentioned shortly in paragraph 9.3.

A condition that is especially detrimental for the fatigue behaviour is if a major part of the fibres is oriented at an angle to the loading direction, as is the case for the BL type specimen, see Fig. 4.11. As the fibre orientation generally is parallel to the mould flow direction MFD, the first general rule is:

1   The loading direction must not be at an angle (of approximately 45°) to MFD, but preferably parallel or perpendicular to this. This is the case *especially in Fatigue loading*.

The consideration a designer has to make in an early stage is as what the main loading of the part will be, one or more directional. If the loading will be in one direction only, a high degree of anisotropy is desirable. If loading in more directions occurs, it is desirable to have low anisotropy. The thicknesses of the core and shell layers is what mainly controls this. Generally the fibre orientation in the shell is parallel to MFD, while in the core layer it is perpendicular to MFD. This leads to the second rule:

2   If the thicknesses of core and shell layer are approximately equal, the anisotropy of mechanical properties is low. Thick shell layer and thin core lead to a high strength parallel to MFD, while a thin shell and thick core lead to the opposite, a higher strength perpendicular to MFD.

The orientation in the shell layers is generally parallel to MFD. The orientation in the core layer is perpendicular to MFD far from the sides of the mould, and more parallel to MFD close to the sides. This leads to the third rule:

3   Thick shell layers give low inhomogeneity of properties, while a thick core layer results in more variation of properties from place to place in the final product.

Shear flow leads to orientation in MFD, close to the gate the material has experienced more shear, and should be more highly oriented in flow direction [Hegler,84]. However, an increase in fibre alignment with flow path has also been reported, see §6.3.2.2 and [Bright,78]. Thus it is impossible to directly correlate the degree of fibre orientation with distance to the gate. However, at the end of the flowpath the alignment of the fibres in the shell layers changes to perpendicular to MFD.

4   Near the end of the flow path, the orientation in the shell layers changes to perpendicular to MFD.

The last rule results from the orientation effect of diverging and converging flows, respectively resulting in orientation perpendicular to and parallel to MFD.

5   A decrease in thickness or cross section of the part, as seen in the direction of the flow path, leads to an increase in orientation in MFD. On the other hand an increase in thickness, leads to a decrease in fibre orientation. If locally a higher load bearing capacity of the part is needed, it is generally *not possible* to accomplish this by an increase in wall thickness.

Chapter 9, Practical applications of the research   147

The last consideration on this subject is the existence of weld lines, which occur if two melt fronts meet. Hot and cold welds are distinguished: Hot welds occur when the flow front is separated by a small obstacle, for making a small hole in the final product. After the obstacle the flows join and continue to fill the mould (Fig. 9.8). Cold welds occur when the melt has been separated over a long flowpath, see Fig. 9.9. Often the flow fronts meet head-on at a cold weld. Weld lines in fibre reinforced polymers behave differently as compared to not reinforced polymers. Where for not reinforced polymers the strength of the weld line increases with increasing distance from the point where the melts met (for a hot weld), this is not the case for reinforced plastics. Fibres are aligned along the weld line, with no fibres crossing. Therefore the strength of the weld line can be as low as 30 - 65% of the strength without weld line [Akay,93]. Therefore the design of the part and mould must avoid severe loading of weld lines.

A positive consequence of weld lines is that, the fibres being oriented along the weld line in the case of a hot weld. The strength *along the weld line* is higher than for the case of no weld line.

The implications of these rules on the design of the part and mould will be discussed in the following paragraph.

9.3   Mould design

In this chapter some of the basic aspects of fibre orientation, as resulting from the mould design are illustrated. Simple square plates are shown as example products. The loading on this part (the main loading) is indicated by a dot and arrow, indicating the direction of loading only (Load is not applied at the dot only!). The location of the gate(s) is (are) indicated by arrows with a perpendicular line. Finally holes in the product are indicated by hatched squares (for simplicity all holes are drawn square). A good source for examples of fibre orientation in injection moulded parts is Hegler, 1984.

According to rule 1 the fibres must not be at an angle to the load for good fatigue behaviour, as in Fig. 9.1 (left). An improved mould design is shown on the right, where the position of the gate is changed in order to align MFD with the loading direction. However, this will create orientation perpendicular to MFD, due to the diverging flow, and is therefore not the optimal design.

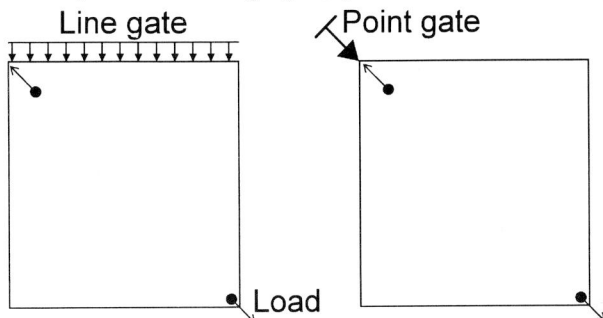

Figure 9.1   Left: Loading at an angle to MFD. According to rule 1 only, the solution to the right is preferable.

The design of the part, from the view of fibre orientation, begins with the consideration if the part needs to have more orthotropic or anisotropic properties (Rule 2). In Fig. 9.2 this is shown schematically. In the case of bi-axial loading, more isotropic properties will be desirable. In this case the thicknesses of both shell layers and the core layer should be approximately equal.
To obtain anisotropic properties, with high strength in the flow direction, the shell layer should be thick compared with the core layer.

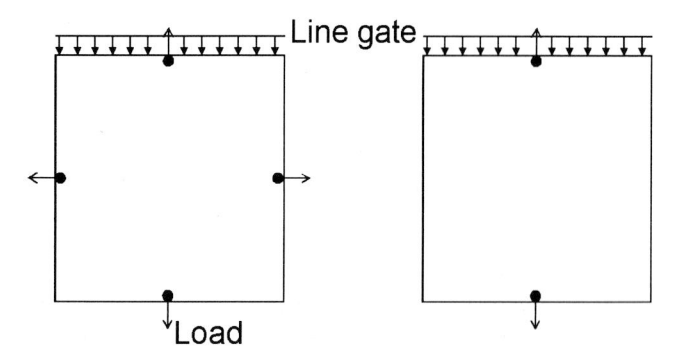

Figure 9.2   Loading of a short fibre reinforced plastic plate. Left: othotropic properties are needed, right: anisotropic properties can be preferable.

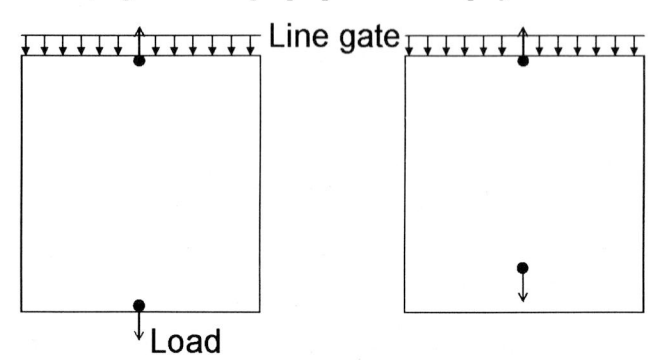

Figure 9.3   The orientation towards the end of the flowpath changes from aligned with MFD towards perpendicular to MFD. Therefore the solution on the right is preferable.

The orientation of the fibres towards the end of the flowpath changes, with less orientation in flow direction. Therefore it is not preferred to exert a load on the part at the end of the flowpath, as shown on the left picture in Fig. 9.3. On the right an improved solution is shown.

I the main loading direction must be, for some reason, perpendicular to MFD, it is preferable to put the load as far from the gate as possible, rule 4, Fig. 9.4.

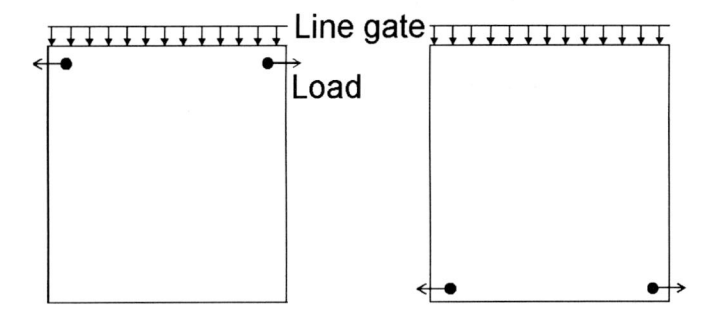

Figure 9.4   *The orientation towards the end of the flowpath changes from aligned with MFD towards perpendicular to MFD. Therefore the solution on the right is preferable.*

Rule 5 indicates the influence of increasing or decreasing cross section with mould flow direction. If the cross section decreases with MFD, the orientation of the fibres will increase, and the orientation of the fibres will become perpendicular to MFD with decreasing cross section. In Fig. 9.5 this is illustrated.

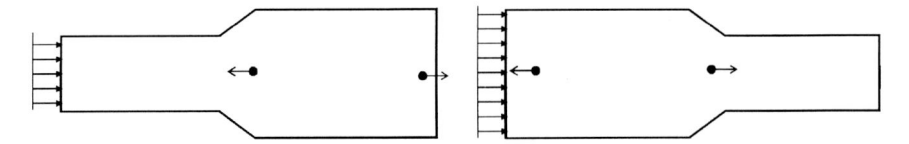

Figure 9.5   *The orientation decreases with diverging flow (increasing cross section), therefore the solution on the right is preferable.*

Changes of thickness have a similar effect as shown in Fig. 9.5, increasing the thickness will, contrary to classic engineering materials, not necessarily increase the load bearing capacity.

The examples shown above assume a *tensile loading* of the product. In that case all layers through the thickness are equally strained, implying that all layers are equally important in the stress build-up over the thickness of the part. When a product is loaded *in bending* however, the outer layers are strained more than the centre layers. Therefore in the case of bending the first condition is, that in the highly strained layers, the outer layers, the fibres must be parallel to the straining direction. The effect of the differences in tensile and bending loading will be illustrated for the example shown in Fig. 9.2. For the case of loading in both directions, on the left, the shell layer should be thin, as it is on the outer layer it will provide sufficient stiffness and strength for bending in the Mould Flow Direction. The core layer should be relatively thick, in order to get fibres with orientation in the effective direction, as far from the neutral line as possible. This is different from the optimal layer distribution for the tensile case, where a similar thickness for shell layers and core layer is required. In Figure 9.6 both cases are compared. If the bending loading is in only one direction, as in Fig. 9.2 on the right, the shell layers should be as thick as possible. This is the same as in the case of tensile loading.

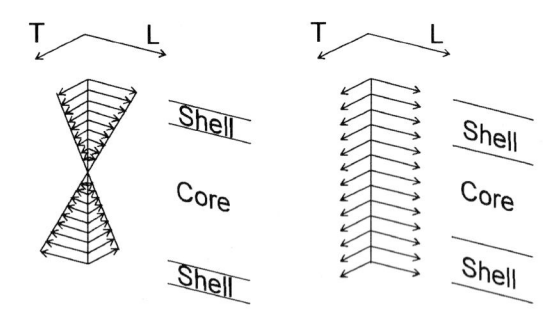

Figure 9.6 *Strain distributions through thickness in L and T direction (respectively parallel and perpendicular to MFD) due to bending loading in both directions (left) and bi-axial tensile loading (right). Optimum layer thicknesses are illustrated.*

The last subject discussed in paragraph 9.2 is weld lines. Weld lines are weaknesses in the product, due to the joining of polymer flows from different directions, and loading of these must be avoided. A distinction must be made between cold and hot welds. In the case of a hot weld, the melts are separated over only a short distance, for example around a small hole in the part, Fig. 9.7. For unreinforced plastics this is the best case, because the melts are relatively hot, and mixing of these will generally take place. However, in the case of reinforced plastics, even hot welds signify a strong decrease in strength, because fibres are oriented parallel to the weld, with no fibres crossing the weld. A cold weld, see Fig. 9.8, appears if melts are separated over a long flowpath, for example when moulding a ring-like structure. Even for not reinforced plastics this type of weld brings about a strong decrease in strength, as the melt is relatively cold, and mixing of both melts hardly takes place. For reinforced plastics this effect is worsened by the fact that no fibres cross the weld line.

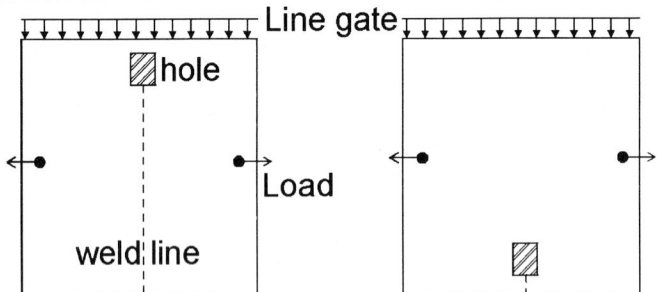

Figure 9.7 *Generally the loading of weld lines must be avoided, because of the low strength of these. Therefore the solution on the right is preferable.*

Generally it must be avoided that weld lines are severely loaded, as far as this is possible in the design of the product and the mould (Fig. 9.7). The location of the weld line can be influenced by changing the design of the mould, generally the position of the gates is crucial.

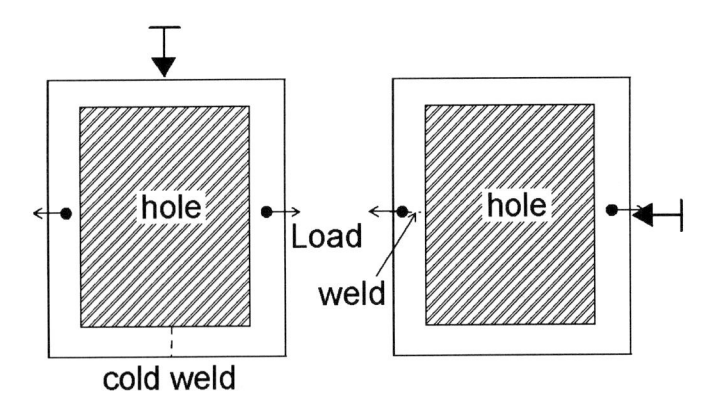

Chapter 9, Practical applications of the research   151

Figure 9.8   *The cold weld shown here can not be severely loaded. On the right a possible solution is shown, depending on how the load is applied.*

In the case of cold welds the situation is still more complicated. If the square ring as shown in Fig. 9.8 is loaded, the position of the weld line can be only shifted, not generally to an area without severe loading. One (expensive) solution is push-pull injection moulding (Fig. 9.9).

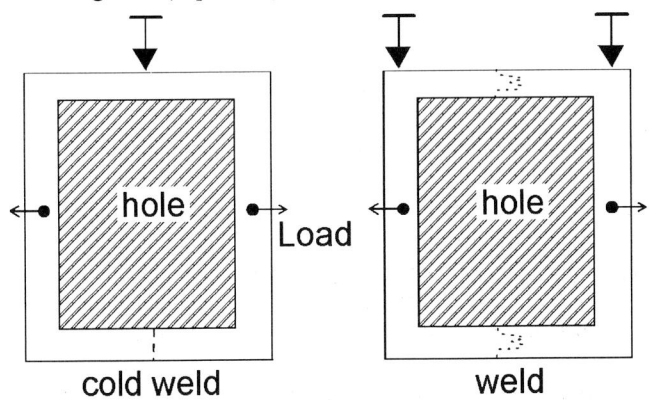

Figure 9.9   *A technical solution to avoid weld lines, as shown on the left, exists in push-pull moulding. On the right the position of the **two** gates is shown.*

In the case of push-pull moulding the mould must be equipped with two gates [Ludwig,95, Becker,93]. During filling, the mould is filled through both gates. When the mould is filled, more material is injected through one gate, *while material leaves the mould through the other gate*. Consequently the melt can be returned, by reversing this process in the other direction. This can be repeated a number of times. The effect is that the material at the weld lines is mixed, increasing the weld line strength. Furthermore, due to the extra movement of the melt, more shear flow is created, leading to a higher fibre orientation in the Mould Flow Direction.

As was stated earlier fibres near the weld lines do have an orientation parallel to the weld line: The strength and stiffness along the weld line is higher than without weld line, because of the absence of a core layer, Fig. 9.10. Obviously the transverse strength is affected negatively by the multiple weld lines that are introduced using a multipoint gate.

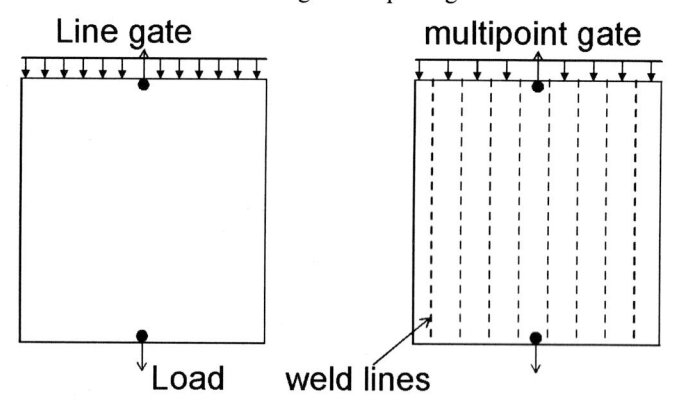

Figure 9.10   Fibre orientation is always along the weld line, reason why the example on the right, using a multipoint gate, introducing weld lines, gives a higher strength in flow direction. [Geerling,93]

As was seen in this last example, of influence is not only the design of the mould, but also of the runner system. The orientation the fibres have when entering the mould is influenced by the position, shape and dimensions of the runner system.

9.4   Influence of processing conditions

The influence of processing conditions is a complex one, and may influence more than the fibre orientation alone. Apart from the fibre orientation the processing conditions will influence the degree of fibre breakage during injection moulding, with more fibres breaking at higher injection speeds and higher matrix viscosity (low temperature of melt and mould). The processing conditions may also influence the matrix cristallinity and spherulite size.

To some extent the fibre orientation can be influenced by the injection moulding parameters, especially injection speed and melt respectively mould temperature. Generally the effect of increasing injection speed is an increase in core layer thickness [Akay,91, Karger-Kocsis,90, Bay,89]. This is caused by the increase in shear thinning, leading to a more plug-like flow, Fig. 2.3. Sometimes the same effect (increasing core thickness) is seen for increasing temperature of melt and mould [Akay,91].

Concluding: Both a low injection speed as well as a high viscosity lead to a thick shell, and more fibre orientation in mould flow direction. However, at low injection speed less fibre fracture will occur, while lowering the temperature will increase fibre fracture during injection moulding.

9.5   Fatigue Data

The practical use of these methods to obtain the fatigue behaviour of parts, requires the availability of fatigue data. As was shown in chapter 4, the position of the Master Curve is dependent on water content, fibre content, fibre aspect ratio and of course the matrix grade and the type of fibres. Thus the Master Curve must be established, using the same material as in the part that is to be designed. The tensile strength must be measured, and the fatigue lifetime must be established. The load levels at which this must be done can be adapted to the purpose of the part; the number of cycles to failure required. To establish the Master Curve correctly, all experiments must be done repeatedly, at least 3, but ideally 5 experiments per load level.

As the cyclic creep, presented in Chapter 4, gives a good measure of the increase of damage in the specimen, the cyclic creep provides us with a method to establish the *fatigue limit*. The fatigue limit is the fatigue load level at which the fatigue life is infinite. Especially for metals the fatigue limit is a realistic value, while for plastics and for fibre reinforced plastics the slope of the Wöhler curve decreases with decreasing fatigue stress, but is not horizontal. Practically the experiment could be as follows: The specimen should be loaded with a fatigue load at which no fatigue damage will be created, e.g. at 25% of UTS. The development of the cyclic creep must then be evaluated over a sufficiently long period of time, at least 10^6 cycles. If the cyclic creep speed equals zero, the predicted lifetime for the specimen at that load is infinite. Increasing the fatigue load in small steps will lead to a load level where creep can be observed, which is the load exactly above the fatigue limit. However, care must be taken that no elongation of the specimen due to other causes can take place: The temperature must be exactly constant, and the conditioning of the specimen must be thus that the specimen is in absolute equilibrium. For example for PolyAmide no water must be absorbed during the experiment, as this will lead to a decrease in Modulus and an increase in elongation, which will be indistinguishable from creep.

9.6 Conclusions

The Master Curves as presented first in Chapter 4, can be used to evaluate the design of a part, as far as the fatigue performance is concerned. In this chapter some general indications are given of what fibre orientations are required for optimum use of this reinforcement. Some basic principles are given, and these are illustrated using a simple part. The fibre orientation in a part can be influenced most effectively by changing the location of the injection *gates*. Fibre orientation can be influenced by the injection process parameters to a much lesser extent, apart from the fact that these parameters are often dictated by other considerations, e.g. appearance of the moulded part, degradation of the melt and cost.

Simulation of the fibre orientation becomes more and more reliable. Applying the design rules for short fibre reinforced plastics and simulation of the fibre orientation and fatigue behaviour is very important.
Tests on products should only be carried out to proof the reliability rather than as a "trial and error" method to design or adapt products. Fatigue tests on injection moulded products have to be carried out in order to be sure about the fibre orientation field and weld lines. Fatigue tests of course have the drawback that they take a lot of time, especially with PolyAmide because the energy dissipation is high, and therefore unacceptable temperature rises in the specimen occurs at high testing frequencies (§ 4.3.2). For evaluation purposes fatigue tests at "high" load levels can be done to generate the higher part of the S-N curve.

In this chapter only one aspect of the part design is highlighted, the (fatigue) strength of the part as it is influenced by fibre orientation. Of course various other aspects as shrinkage, warpage etc. can be of prime importance in mould design.

If, although part-design and mould-design have been executed as meticulously as possible, the part does not meet the requirements, mould redesign and changing of the mould will be extremely expensive. In this case it can be possible to use carbon fibre reinforced PolyAmide instead of glassfibre reinforced. The carbon grade is more expensive, but also possesses better properties. The improved properties can be sufficient to meet the requirements, without having to change the mould.

Summary

Fatigue failure is an important design criterium for products produced from fibre reinforced plastic. Products made from this material are being increasingly used in load bearing applications, for example in under-bonnet use in automobiles. However, the fibres in the products are oriented by the melt flow during filling, and therefore these parts have high anisotropy and inhomogeneity of (mechanical) properties. The material is structured in layers, with per layer a certain average fibre orientation. Skin, shell and core layers can be distinguished, with orientations usually random in the plane, aligned with Mould Flow Direction (MFD) and perpendicular to MFD respectively. Degree of orientation in these layers as well as the thicknesses of the layers vary from location to location in the plate.

The main objective of this research was to relate the fatigue behaviour of glassfibre reinforced PolyAmide 6, with the glassfibre orientation. Lifetime experiments were the main interest, although some crack propagation experiments have been executed as well. The varying fibre orientation was obtained by using milled specimens from injection moulded square plates. Specimens cut from different locations in the plate, and with different specimen orientations relative to the mould flow direction, have a different fibre orientation distribution.

On these specimens tensile tests were done, as well as fatigue lifetime tests. The fatigue crack propagation (FCP) tests were executed on centre notched specimens from the same plates. For better understanding of the results and to be able to indicate possible methods to improve the material (fatigue) behaviour, also the failure mechanism in fatigue and in tensile fracture was investigated.

One further complication is that the specific plastic used in this research, PolyAmide 6, is hygroscopic. It absorbs water, which leads to changes in properties like Elastic Modulus, strength and ductility. Therefore the conditioning of specimens is extremely important: Conditioned (in air of 23°C and 50%RH) specimens were the main interest, but specimens that were dry as moulded and specimens which were wet (by emerging these in water) have been investigated as well.

Both for lifetime experiments and for FCP experiments the fatigue strength and the tensile strength show a correlation. For specimens with different fibre orientation, and consequently different properties, the fatigue lifetime is the same, when fatigued at the *same fraction of tensile strength* UTS. The tensile strength of these specimens can vary between 100 and 150MPa, depending on the fibre orientation. The correlation exists for conditioned specimens, as well as for dry as moulded and wet specimens. For the designer this is a very valuable fact, as he must only know, or predict, the tensile strength of the composite with a given fibre orientation, to also know the fatigue behaviour. The fatigue behaviour can be represented in a Master Curve, in which the fatigue lifetime is plotted against the fatigue stress *divided by* the tensile strength. This Master Curve is *independent of* the glassfibre orientation. For different conditions as far as the water absorption is concerned, different Master Curves are valid. Also differences in material cause a shift of the Master Curve. Improved fibre - matrix bonding leads to a longer fatigue lifetime, as is the case for increased fibre length. That the Master Curve exhibits a shift, indicates that these factors do not influence tensile strength and fatigue strength to the same degree.

The fatigue lifetime tests were monitored: displacement, force and temperature were continuously measured, enabling the calculation of the Modulus of elasticity, creep during fatigue and the energy dissipation per load cycle. Especially the creep in fatigue is a good measure for the damage in the specimen. If the creep increase is rapid (a high creep speed) the increase of damage in the specimen is high and the specimen fails in relatively few load cycles. The correlation that exists between the creep speed (which can be measured during the experiment) and the lifetime of the specimen, enables an accurate prediction of the lifetime of a specimen, *during the fatigue experiment*. This reduces the influence of lifetime-scatter on the estimated percentage of lifetime at which an experiment is stopped. Fatigue experiments are stopped before failure to measure the damage inside the specimen, and especially the location of the damage in the thickness. The method that was used to measure damage was microtoming thin slices from the fatigued specimens, and tensile test these. Damage in the specimens could be demonstrated by a decrease in strength of the slices, usually accompanied by an increase in fracture strain. This decrease in strength was significant after 80% of the predicted lifetime. Damage was present especially in the layers with orientation of the glassfibres in loading direction. A very remarkable observation that was made during these experiments was, that no slices had strength zero. This signifies that *no cracks* were present until just before the end of the fatigue process. This although stress whitening zones on the fatigued specimens could be easily observed with the bare eye. Slices containing these stress whitening zones did show an extremely low strength and low fracture strain however, indicating that severe damage was present.

These observations indicated that the fatigue failure mechanism for conditioned (to equilibrium water content) and wet specimens is one of formation and growth of *bridged cracks*. During fatigue damage is formed at the location of highest stress, the fibre ends, where consequently voids develop, accompanied by fibre-matrix debonding. These voids grow together into cracks. However, the crack walls remain bridged by matrix material and/or fibres. The bridged crack grows until the residual strength of the specimen is as low as the maximum fatigue stress, when the specimen fails in the last load cycle. The failure mechanism in the case of dry as moulded material is different. Matrix ductility is much lower, therefore the void formation is less pronounced, and no bridges are allowed to form. So matrix cracking does occur, not matrix plastic deformation. This latter case is the failure mechanism mostly presented in literature. The failure mechanisms could be confirmed and supported using fractography. The Scanning Electron Microscope (SEM) was used to examine the fracture surfaces, where the matrix ductility, fibre pull - out and fibre - matrix bonding can be observed. A technique of cryogenically breaking fatigued specimens was used to reveal the damage inside the material. Fatigued specimens were emerged in liquid nitrogen for at least 5 minutes, and then broken. The resulting fracture surface from this is brittle, consequently all ductility visible on the surface is the result of the fatigue process. Thus bridges could be revealed, as well as voids at fibre ends.

The matrix ductility that gives rise to the formation of voids during the fatigue failure process, must be accompanied by lateral contraction of the matrix. This lateral contraction can impose a tensile stress on the fibre - matrix interface perpendicular to the loading direction, for fibres parallel to the loading direction. This is apart from the shear stress which transfers the load from the matrix to the fibre. Micromechanical Finite Element modelling of this situation has been executed; a high tensile stress on the fibre - matrix interface was found close to the fibre end. For high strains this tensile stress on the interface is higher than the shear stress. This tensile stress shall accelerate fibre - matrix debonding considerably.

Very interesting is the difference between the fracture surfaces of tensile tests and of fatigue tests. In the former the fibre - matrix bonding is *not broken*, the matrix fails, at a short distance from the fibre. In fatigue the fibre - matrix bond itself is broken. Consequence of this different interfacial failure site is, that when the fibre - matrix bonding is improved, e.g. by using an improved coating of the fibres, no effect is seen in tensile properties. To see the effect of the increased bond strength, fatigue experiments are more suitable.

The differences between the various failure mechanisms, in tensile fracture and in fatigue fracture for dry as moulded respectively conditioned material, indicate that the fracture mechanism is not an explanation for the correlation of fatigue strength and tensile strength. The explication of this correlation is in the stress distribution over the thickness of the specimen. The stress build-up over the different layers is the same in fatigue as well as in the tensile test, although at a different stress level.

Finally design rules are given, combined with examples of mould design. The fibre orientation can be influenced mainly by the position of the gate(s) in the mould, but also by the design of the part and of the runner system. Some influence on the fibre orientation can be also had by the injection moulding parameters, like injection time (equivalent with injection speed or injection pressure) and melt and mould temperatures.

Samenvatting

Het voorkomen van falen in vermoeiing is een belangrijk ontwerp criterium voor produkten die gemaakt zijn van glasvezelversterkte plastic. Produkten die van dit materiaal gemaakt zijn, worden meer en meer gebruikt in toepassingen waar ze mechanisch zwaar belast worden, bijvoorbeeld in toepassingen in automobielen. Tijdens het spuitgietproces worden de vezels gericht door de polymeer-stroom, met als gevolg een hoge anisotropie en inhomogeniteit van de (mechanische) eigenschappen van het materiaal in het produkt. Het materiaal krijgt een lagenstructuur, met per laag een bepaalde gemiddelde vezelorientatie. Skin, shell en core lagen worden onderscheiden, met doorgaans als orientatie respectievelijk random, evenwijdig met de stroomrichting (MFD) en loodrecht op de stroomrichting. De mate van orientatie in deze lagen alswel de dikte van de lagen variëren over de plaat.

Het hoofddoel van het onderzoek was om een relatie te vinden tussen het vermoeiingsgedrag van glasvezelversterkt Polyamide 6 met de vezelorientatie. Levensduur experimenten waren hierbij het voornaamste middel, maar er zijn ook scheurgroeiexperimenten uitgevoerd. De variaties in vezelorientatie zijn verkregen door uit vierkante, gespuitgiete platen proefstukken te frezen. Proefstukken die uit verschillende posities van de plaat gehaald zijn, of met een andere proefstukorientatie ten opzichte van de vloeirichting, hebben een verschillende vezelorientatie distributie. Op deze proefstukken zijn trekproeven uitgevoerd, alsmede levensduur experimenten. Vermoeiings scheurgroei-experimenten (FCP) zijn uitgevoerd op proefstukken met een centrale kerf, gemaakt uit dezelfde platen. Ten einde de resultaten beter te begrijpen, en om mogelijke methoden aan te kunnen geven om het materiaalgedrag (in vermoeiing) te verbeteren, is het faalmechanisme in vermoeiing zowel als in een trekproef onderzocht.

Een verdere complicatie bij het specifieke plastic wat in dit onderzoek is gebruikt, Polyamide 6, is dat het hygroscopisch is. Het absorbeert water, hetgeen eigenschappen zoals stijfheidsmodulus, sterkte en ductiliteit verandert. Daarom is de conditionering van de proefstukken uitermate belangrijk: Bij dit onderzoek zijn voornamelijk proefstukken die aan de lucht zijn geconditioneerd gebruikt, maar ter vergelijking zijn ook droge en natte proefstukken onderzocht.

Zowel voor de levensduur experimenten alsook voor de FCP experimenten is een correlatie gevonden tussen de vermoeiingssterkte en de treksterkte. Proefstukken met verschillende vezel oriëntatie, en daardoor verschillende eigenschappen vertonen de zelfde levensduur wanneer ze getest zijn op *dezelfde fractie van de treksterkte* (UTS). Hierbij kan de treksterkte variëren van 100 tot 150MPa, afhankelijk van de vezelorientatie. De correlatie is aangetoond voor geconditioneerde proefstukken, maar ook voor droge en natte proefstukken. Voor de ontwerper is dit een erg bruikbaar gegeven, omdat hij alleen maar de treksterkte hoeft te kennen (of kunnen voorspellen) van de composiet met een bepaalde vezel oriëntatie, om het vermoeiingsgedrag te kennen. Vermoeiingsgedrag kan worden weergegeven met een Master Curve, waarin de levensduur is uitgezet tegen de vermoeiingsspanning *gedeeld door* de treksterkte. Deze Master Curve is *onafhankelijk* van de glasvezeloriëntatie, voor de door ons gemeten orientaties. Voor verschillende condities wat betreft vochtopname gelden verschillende Master Curves. Ook veranderingen in het materiaal verschuiven de ligging van de Master Curve, verbeterde hechting van de vezels aan de matrix leidt tot een langere levensduur, net als langere glasvezels. Dat de Master Curve verschuift geeft dus aan dat de invloed van deze factoren niet hetzelfde is in trek en in vermoeiing.

Tijdens de vermoeiingslevensduur experimenten werden verplaatsing, kracht en temperatuur continu gemeten. Hierdoor was het mogelijk om de elasticiteitsmodulus, de kruip tijdens vermoeiing en de energiedissipatie per belastingscyclus te berekenen. Vooral de kruip tijdens vermoeiing is een goede maat voor de schade in het proefstuk. Als de kruip snel toeneemt (een hoge kruipsnelheid), dan is de toename van de schade in het proefstuk ook snel, en zal het proefstuk binnen een relatief klein aantal cycli falen. De correlatie die bestaat tussen de kruipsnelheid (welke tijdens het experiment gemeten kan worden) en de levensduur van het proefstuk, maakt het mogelijk om een nauwkeurige voorspelling te doen van de levensduur van het proefstuk, *tijdens het experiment*. Dit vermindert de invloed die de spreiding van de levensduur heeft op het geschatte percentage van de levensduur waarbij een vermoeiingsexperiment wordt gestopt. Deze experimenten worden gestopt voordat het proefstuk bezweken is, om de schade in het proefstuk te meten, en vooral de positie van die schade in de dikte. De methode die hiervoor is gebruikt is het microtomeren van dunne plakjes van het proefstuk, welke vervolgens een trekproef ondergingen. Schade in het proefstuk blijkt uit een afname van de sterkte van de schijfjes, meestal vergezeld van een toename in rek tot breuk. Na 80% van de voorspelde levensduur is de gemeten daling van de sterkte significant. Schade was aanwezig vooral in de lagen waar de vezels in belastingsrichting georiënteerd zijn. Een erg opmerkelijke waarneming tijdens deze experimenten was, dat geen schijfjes met sterkte nul gevonden zijn. Er waren dus *geen scheuren* aanwezig aan het eind van het vermoeiingsproces. Dit ondanks dat er "stress whitening zones" met het blote oog zichtbaar waren op de vermoeide proefstukken. Schijfjes waarin deze "stress whitening zones" zichtbaar waren, vertoonden wel een buitengewoon lage sterkte en lage rek tot breuk, hetgeen aangeeft dat ernstige schade aanwezig was.

Deze waarnemingen duiden op een vermoeiingsmechanisme voor *geconditioneerde* en *natte* proefstukken van ontstaan en groeien van *overbrugde scheuren*. Tijdens vermoeiing ontstaat schade aan de vezeluiteinden, waar de spanningsintensiteit het hoogst is. Vanuit deze schade groeien holten, terwijl de vezel uit de matrix onthecht. De holten groeien samen tot scheurtjes, maar de scheurwanden blijven overbrugd door matrix materiaal en/of vezels. Deze overbrugde scheur groeit totdat het resterend draagvermogen gezakt is tot de vermoeiingsbelasting, en het proefstuk faalt dan in de laatste belastingscyclus. Het mechanisme voor *droog* materiaal verschilt hiervan: De matrix ductiliteit is veel minder, waardoor de holtevorming praktisch afwezig is, en geen bruggen gevormd kunnen worden. Er vindt dus meer breuk plaats van de matrix, dan plastische vervorming. Dit laatste breukmechanisme in vermoeiing wordt meestal beschreven in de literatuur. Van beide breukmechanismen kon het bestaan worden bevestigd met behulp van fractografie. Scanning Electron Microscopie (SEM) is gebruikt om de breukvlakken te bestuderen, waarop de ductiliteit van de matrix, vezel pull-out en vezel - matrix hechting kon worden gezien. Een techniek die gebruikt is om de schade in het materiaal bloot te leggen was het cryogeen breken van vermoeide proefstukken. Vermoeide proefstukken werden ondergedompeld in vloeibaar stikstof voor tenminste 5 minuten en vervolgens gebroken. Omdat het resulterende breukvlak hiervan bros is, moet alle ductiliteit die zichtbaar is het gevolg zijn van het vermoeiingsproces. Op deze manier konden bruggen worden aangetoond, alsmede de holten aan de vezeluiteinden.

De ductiliteit van de matrix is aanleiding voor de vorming van holten tijdens het vermoeiingsproces. Deze plastische vervorming van de matrix moet worden vergezeld door dwarscontractie. Deze dwarscontractie wordt verhinderd door de hechting van de matrix aan de vezels parallel aan de belastingsrichting, en kan daardoor leiden tot een trekspanning op de interface, loodrecht op de richting van belasten. Met behulp van eindige elementen methoden is een micromechanisch model van deze situatie gemaakt. Op deze wijze is aangetoond dat een hoge trekspanning op het interface aanwezig is, vlakbij het vezeluiteinde. Bij een hoge rek is deze trekspanning op het interface groter dan de schuifspanning, en zal daarom de onthechting van de vezel sterk versnellen.

Erg interessant is het verschil tussen de breukvlakken van de trekproeven en van de vermoeiingsexperimenten. In het eerste geval wordt de vezel - matrix hechting *niet verbroken*, maar de matrix zelf bezwijkt, op een kleine afstand van de vezel. In vermoeiing daarentegen wordt deze hechting wèl verbroken. Gevolg hiervan is dat, wanneer de hechting tussen vezel en matrix wordt verbeterd, bijvoorbeeld door het gebruiken van een verbeterde coating van de vezel, er geen effect is op de treksterkte. Om het effect van de verbeterde hechting te kunnen zien, zijn vermoeiingsexperimenten beter geschikt.

De verschillen tussen de breukmechanismen, van trekproef en vermoeiing van droog respectievelijk geconditioneerd materiaal, geeft aan dat het breukmechanisme geen verklaring is voor de correlatie van vermoeiingssterkte en treksterkte. De verklaring van deze correlatie ligt in de spanningsverdeling over de dikte van het proefstuk. De spanningsopbouw over de verschillende lagen is hetzelfde in vermoeiing als in een trekproef, alleen bij een lager niveau van de spanning.

Ontwerpregels wat betreft de vezeloriëntatie zijn gegeven, alsmede voorbeelden van matrijsontwerpen. De vezeloriëntatie kan worden beïnvloed voornamelijk door de positie van de aanspuiting(en) in de matrijs, maar ook door het ontwerp van het produkt en het runner systeem. Een kleine invloed op de vezeloriëntatie kan ook worden uitgeoefend door de spuitgiet-parameters, zoals injectietijd (of injectiesnelheid of injectiedruk) en de matrijs en smelt temperaturen.

References

Adkins,88	**Adkins D.W., Kander R.G.**, Fatigue performance of glass reinforced thermoplastics. How to apply advanced composites technology, *Proceedings of the fourth annual conference on advanced composites*, 13-15 september 1988, pp. 437-445.
Advani,87	**Advani S.G., Tucker III C.L.**, The use of tensors to describe and predict fiber orientation in short fiber composites. *J. of rheol.* **31**, 1987, pp. 751-784.
Agarwal,90	**Agarwal B.D., Broutman L.J.**, Analysis and performance of fiber composites. John Wiley & Sons, 1990.
Akay,91	**Akay M., Barkley D.**, Fibre orientation and mechanical behaviour in reinforced thermoplastic injection mouldings. *J. Mater. Sci.* **26**, 1991, pp. 2731-2742.
Akay,94	**Akay M.**, Moisture Absorption and its Influence on the Tensile Properties of Glass-Fibre Reinforced Polyamide 6,6. *Pol. & Pol. Comp.* **2**, 1994, pp. 349-354.
Altan,90	**Altan M.C., Subbiah S., Güçeri S.I., Byron Pipes R.**, Numerical Prediction of Three-Dimensional Fiber Orientation in Hele-Shaw Flows. *Pol. Eng. Sci.* **30**, 1990, pp.848-859.
Baraldi,92	**Baraldi U., Lucarelli J.F., Fouarge A.**, Fibre orientation Prediction in Injection moulded reinforced thermoplastics products. *Pol. Eng.* **11**, 1992, pp.1-38.
Bay,89	**Bay R.S., Tucker III C.L., Davis** R.B., Effects of processing on fiber orientation in simple injection moldings. *Annual technical conference-society of plastics engineers, 1-4 may 1989, ANTEC 89*, pp. 539-542.
Bay,91a	**Bay R.S., Tucker III C.L.**, Fiber orientation in simple injection moldings: part 1 - theory and numerical methods. *MD-Vol 29, Plastics and plastic composites: material properties, part performance and process simulation, ASME 1991*, pp. 445-471.
Bay,91b	**Bay R.S., Tucker III C.L.**, Fiber orientation in simple injection moldings: part 2 - experimental results. *MD-Vol 29, Plastics and plastic composites: material properties, part performance and process simulation, ASME 1991*, pp. 473-492.

Bay,92	**Bay R.S., Tucker III C.L.**, Stereological measurements and Error Estimates for Three-Dimensional Fibre Orientation. *Pol. Eng. Sci.* **32**, 1992, pp. 240-253.
Becker,93	**Becker H., Fischer G., Müller U.**, Push-pull Injection Moulding of Industrial Products. *Kunststoffe* **83**, 1993, pp. 3-4
Bohse,92	**Bohse J.**, Phasenhaftung und Schallemision bei faserverstärkten Thermoplasten. *Kunststoffe* **82**, 1992, pp. 72-76.
Bright,78	**Bright P.F., Crowson R.J., Folkes M.J.**, A study of the effect of injection speed on fibre orientation in simple mouldings of short glass fibre-filled polypropylene. *J. Mater. Sci.* **13**, 1978, pp. 2497-2506.
Bright,81	**Bright P.F., Darlington M.W.**, Factors influencing fibre orientation and mechanical properties in fibre reinforced thermoplastics injection mouldings. *Plast. Rub. Proc. Appl.* **1**, 1981, pp. 139-147.
Brockmüller,92	**Brockmüller K.M., Friedrich K.**, Elastoplastic stress analysis of a short fibre reinforced composite using a three dimensional finite element model with several close to reality features. *J. Mater. Sci.* **27**, 1992, pp. 6506-6512.
Brockmüller,95	**Brockmüller K.M., Bernhardi O., Maier M.**, Determination of fracture stress and strain of highly oriented short fibre-reinforced composites using a fracture mechanics-based iterative finite-element method. *J. Mater. Sci.* **30**, 1995, pp. 481-487.
deBruijn,92	**de Bruijn J.C.M.**, The failure behaviour of high density polyethylene with an embrittled surface layer due to weathering. PhD Thesis, Delft University Press, Delft, 1992.
Busschen,95	**ten Busschen A., Selvadurai A.P.S.**, Mechanics of the Segmentation of an Embedded Fiber, Part I: Experimental Investigations. J. of Appl. Mech. **62**, 1995, pp. 87-97
Carling,85	**Carling M.J., Manson J.A., Hertzberg R.W., Attalla G.**, Effects of fiber orientation and interfacial adhesion on fatigue crack propagation in SGFR polypropylene composites. *Antec 1985, April 29 - may 2 1985*, pp. 396-398.
Cengiz Altan,90	**Cengiz Altan M., Subbiah S., Selçuk I., Güçeri I., Byron Pipes R.**, Numerical Prediction of Three -Dimensional Fiber Orientation in Hele-Shaw Flows. *Pol. Eng. Sci.* **30**, 1990, pp.848-859.

References

Chow,91 — Chow C.L., Lu T.J., Characterization of Fatigue Crack Propagation in Short Fiber Reinforced Thermoplastics - A Unified Approach. *J. Reinf. Plast. Comp.* **10**, 1991, pp. 58-83.

Cox,52 — Cox H.L., The elasticity and strength of paper and other fibrous materials. *Br. J. of Appl. Phys.* **3**, 1952, pp. 72 - 79.

Dally,69 — Dally J.W., Carrillo D.H., Fatigue Behavior of Glass-Fiber Fortified Thermoplastics. *Pol. Eng. Sci.* **9**, 1969, pp. 433 - 444.

Darlington,91 — Darlington M.W., Short fibre reinforced thermoplastics: properties and design. *Durability of polymer based composite systems for structural applications, proceedings of the international colloquium held in Brussels, Belgium, 27-31 August 1990, ed. Cardon A.H. and Verchery G., Elsevier applied science,* 1991, pp. 80-98.

Delpy,85 — Delpy U., Fischer G., Effect of Mold-Filling Conditions on Fiber Distribution in Injection-Molded Disks and on the Mechanical Properties of Such Disks. *Adv. in Pol. Tech.* **5**, 1985, pp.19-26.

Devaux,90 — Devaux E., Chabert B., Non-isothermal crystallization of glass fibre reinforced polypropylene. *Pol. comm.* **31**, 1990, pp. 391-394.

Dibenedetto,79 — Dibenedetto A.T., Salee G., Fatigue crack propagation in Graphite Fiber Reinforced Nylon 66. *Pol. Eng. Sci.* **19**, May 1979, pp. 512-518.

Folgar,84 — Folgar F., Tucker III C.L, Orientation behaviour of Fibers in Concentrated suspensions. *J. Reinf. Plast. Comp.* **3**, 1984, pp. 98-119.

Folkes,80 — Folkes M.J., Russell D.A.M., Orientation effects during the flow of short-fibre reinforced thermoplastics. *Polymer* **21**, 1980, pp. 1252-1258.

Folkes,82 — Folkes M.J., Short Fibre reinforced thermoplastics. Research Studies Press, John Wiley & Sons, 1982.

Frahan,92 — Henry de Frahan H., Verleye V., Dupret F., Crochet M.J., Numerical prediction of fiber orientation in injection molding. *Pol. Eng. Sci.* **32**, 1992, pp. 254-266.

Friedrich,86 — Friedrich K., Walter R., Voss H., Karger-Kocsis J., Effect of short fibre reinforcement on the fatigue crack propagation and fracture of PEEK-matrix composites. *Composites* **17**, 1986, pp. 205-216.

Gadala-Maria,93 — Gadala-Maria F., Parsi F., Measurement of Fiber Orientation in Short-Fiber Composites Using Digital Image Processing. *Pol. Comp.* **14**, 1993, pp. 126-131.

Geerling,93	**Geerling M.P.W.**, Het optimaliseren van ventielhuizen in vacuümrembekrachtigers. Masters Thesis, TUDelft, August 1993.
Hartingsveldt,87	**van Hartingsveldt E.A.A.**, Interfacial adhesion and mechanical properties of Polyamide-6/glass bead composites. Thesis, Delft, 1987.
Hegler,84	**Hegler R.P.**, Faserorientierung beim Verarbeiten kurzfaserverstärkter Thermoplaste. *Kunststoffe* **74**, 1984, pp. 271-277.
Hertzberg,80	**Hertzberg R.W., Manson J.A.**, Fatigue of engineering plastics. Academic Press, 1980.
Hitchen,93a	**Hitchen S.A., Ogin S.l.**, Damage accumulation during the fatigue of an injection moulded glass/nylon composite. *Comp. Sci. Tech.* **47**, 1993, pp. 83-89.
Hitchen,93b	**Hitchen S.A., Ogin S.l.**, Matrix cracking and Modulus reduction during the fatigue of an injection moulded glass/nylon composite. *Comp. Sci. Tech.* **47**, 1993, pp. 239-244.
Hine,93	**Hine P.J. Duckett R.A. Davidson N. Clarke A.R**, Modelling of the elastic properties of fibre reinforced composites. I: Orientation measurement. *Comp. Sci. Tech.* **47**, 1993, pp. 65-73.
Horst,93a	**Horst J.J.**, Literatuurstudie: Mechanische eigenschappen en vermoeiing van kort-glasvezelversterkte thermoplasten. Internal report LMB nr. K-272, TU Delft, February 1993.
Horst,95a	**Horst J.J.**, Determination of Fatigue Damage in short Glassfibre reinforced Polyamide. *Proceedings of the 3rd International Conference on Deformation and Fracture of Composites, 27-29 March 1995, Surrey, UK.* pp. 473-482.
Horst,96a	**Horst J.J., Spoormaker J.L.**, Mechanisms of Fatigue in Short Glass Fiber Reinforced Polyamide 6. *Pol. Eng. Sci.* **36**, 1996, pp. 2718-2726.
Horst,96b	**Horst J.J.**, Fatigue tests on standard and experimental grade glassfibre reinforced PA6. *Joint Research project with DSM. External report K346, TU Delft,* 1996.
Horst,97a	**Horst J.J., Spoormaker J.L.**, Fatigue fracture mechanisms and fractography of short glassfibre reinforced polyamide 6. *J. Mater. Sci.* **32**, 1997, pp. 3641-3651.

Horst,97b	**Horst J.J., Salienko N.V., Spoormaker J.L.**, Fibre - matrix debonding stress analysis for short fibre reinforced materials with matrix plasticity, finite element modelling and experimental verification. *Composites, submitted 1997.*
Jeffery,22	**Jeffery G.B.**, The motion of ellipsoidal particles immersed in a viscous fluid. *Proc. Roy. Soc.* **A102**, 1922, pp. 161.
Jinen,86	**Jinen E.**, Accumulated strain in low cycle fatigue of short carbon-fibre reinforced nylon 6. *J. Mater. Sci.* **21**, 1986, pp. 435-443.
Jinen,89	**Jinen E.**, Influence of fatigue damage on tensile creep properties of short carbon fibre reinforced nylon-6-plastic. *Composites* **20**, 1989, pp. 329-339.
Kalinka,90	**Kalinka G., Boro I., Augustin G., Hampe A., Hinrichsen G.**, Ermüdungsverhalten von Hochleistungs-Faserverbundwerkstoffen mit themoplastischer Matrix. *Kunststoffe* **80**, 1990, pp. 626-630.
Kaliske,73	**Kaliske G., Seifert H.**, Formfüllstudie beim Spritzgießen von glasfaserverstärktem Polyamid-6. *Pl. u. Kaut.* **20**, 1973, pp. 873-840.
Karbhari,89	**Karbhari V.M., Parks B., Dolgopolsky A.**, Effect of mean load levels on fatigue in random short-fibre injection moulded composites. *J. of Mater. Sci. Let.* **8**, 1989, pp. 220-221.
Karbhari,90	**Karbhari V.M., Dolgopolsky A.**, Transitions between micro-brittle and micro-ductile material behaviour during FCP in short-fibre reinforced composites. *Int. J. of Fat.* **12**, 1990, pp. 51-56.
Karger-Kocsis,88	**Karger-Kocsis J., Friedrich K.**, Fatigue crack propagation in short and long fibre-reinforced injection-moulded PA6.6 composites. *Composites* **19**, 1988, pp. 105-114.
Karger-Kocsis,90	**Karger-Kocsis J.**, Effects of processing induced microstructure on the fatigue crack propagation of unfilled and short fibre-reinforced PA-6. *Composites* **21**, 1990, pp. 243-254.
Kelly,65	**Kelly A., Davies G.J.**, The principles of the fibre reinforcement of metals. *Metal. Revs.* **10**, 1965, pp. 2-77.
Konicek,87	**Konicek T.S.**, A method to determine three-dimensional fiber orientation in fiber reinforced polymers. Ms Thesis, University of Illinois at Urbana-Champaign, 1987.

Lang,81	**Lang R.W., Manson J.A., Hertzberg R.W.**, Effects of fibrous and particulate reinforcements on fatigue crack propagation in polyamids. *Org. Coat. Plast. Chem.* **45**, 1981, pp. 778.
Lang,83	**Lang R.W., Manson J.A., Hertzberg R.W.**, Fatigue crack propagation in short-glass-fiber reinforced nylon 66; effect of frequency. *Proceedings of a joint US - Italy symposium on composite materials, June 1981, in Capri, Italy, Plenum Press*, 1983, pp. 377-396.
Lang,84	**Lang R.W., Hahn M.T., Hertzberg R.W., Manson J.A.**, Effects of Specimen Configuration and Frequency on Fatigue Crack Propagation in Nylon 66. *Fracture Mechanics: Fifteenth Symposium, RJ Stanford, Ed., American Society for Testing and Materials*, 1984, pp. 266-283.
Lang,87a	**Lang R.W., Manson J.A.**, Crack tip heating in short-fibre composites under fatigue loading conditions. *J. Mater. Sci.* **22**, 1987, pp. 3576.
Lang,87b	**Lang R.W., Manson J.A., Hertzberg R.W.**, Mechanisms of fatigue in short glass fibre reinforced polymers. *J. Mater. Sci.* **22**, 1987, pp. 4015.
Lees,68a	**Lees J.K.**, A study of the Tensile Modulus of Short Fiber Reinforced Plastics. *Pol. Eng. Sci.* **8**, 1968, pp. 186-194.
Lees,68b	**Lees J.K.** A study of the Tensile Strength of Short Fiber Reinforced Plastics. *Pol. Eng. Sci.* **8**, 1968, pp. 195-201.
Ludwig,95	**Ludwig H.-C., Fischer G., Becker H.**, A quantitative comparison of morphology and fibre orientation in push-pull processed and conventional injection-moulded parts. *Comp. Sci. Tech.* **53**, 1995, pp. 235-239.
Malzahn,84	**Malzahn J.C., Friedrich K.**, Fracture resistance of short glass and carbon fibre/polyamide 6.6 composites. *J. Mater. Sci. Lett.* **3**, 1984, pp. 861-866.
Mandell,80	**Mandell J.F., Huang D.D., McGarry F.J.**, Fatigue of glass and carbon fiber reinforced engineering thermoplastics. *Preprint of the thirty-fifth annual conference, reinforced plastics/ composites institute, New Orleans, Louisiana, February 5-8,* 1980, pp. 20-D 1-11.
Mandell,83	**Mandell J.F., McGarry F.J., Li C.G.**, Fatigue crack growth and lifetime trends in injection molded reinforced thermoplastics. Research report R83-1, Dept. of Mats. Sci. Eng., MIT, Cambridge, Massachsetts.

Matsuoka,90	**Matsuoka T., Takabatake J.I., Inoue Y., Takahashi H.**, Prediction of fiber orientation in injection molded parts of short-fiber-reinforced thermoplastics. *Pol. Eng. Sci.* **30**, 1990, pp. 957-966.
McNally,77	**McNally D.**, Short fiber orientation and its effects on the properties of thermoplastic composite materials. *Pol. Plast. Tech. Eng.* **8**, 1977, pp. 101-154.
McCrum,88	**McCrum N.G., Buckley C.P., Bucknall C.B.**, Principles of Polymer Engineering, Oxford University Press, New York, 1988.
Monette,93	**Monette L., Anderson M.P., Grest G.S.**, Effect of interphase modulus and cohesive energy on the critical aspect ratio in short-fibre composites. *J. Mater. Sci.* **28**, 1993, pp. 79-99.
O'Donnell,94	**O'Donnell B., White J.R.**, Young's modulus variations within short fibre reinforced nylon 6,6 injection mouldings. *Plast., Rub. Comp. Proc. Appl.* **22**, 1994, pp 69-77.
Paterson,92	**Paterson M.W.A., White J.R.**, Effect of water absorption on residual stresses in injection-moulded nylon 6,6. *J. Mater. Sci.* **27**, 1992, pp. 6229-6240.
Piggott,80	**Piggott M.R.**, Load bearing fibre composites. Pergamon press, 1980.
Pipes,82	**Pipes R.B., McCullough R.L., Taggart D.G.**, Behavior of discontinuous fiber composites: Fiber orientation. *Pol. Comp.* **2**, 1982, pp. 34-39.
Ranganathan,90	**Ranganathan S., Advani S.G.**, Characterization of orientation clustering in short-fiber composites. *J. of Pol. Sci.: B: Pol. Phys.* **28**, 1990, pp. 2651-2672.
Remmerswaal,90	**Remmerswaal J.A.M.**, Fatigue of Amorphous polymers. Thesis, Delft, 1990.
Sato,82	**Sato N., Sato S., Kurauchi T.**, Fracture mechanism of short glass fiber reinforced polyamide thermoplastics. *Progress in science and engineering of composites, volume 2, Proceedings of the fourth international Conference on Composite materials, ICCM-IV, October 25-28,* 1982, Tokyo, pp.1061-1066.
Sato,84	**Sato N., Kurauchi T., Sato S., Kamigaito O.**, Mechanism of fracture of short glass fibre-reinforced polyamide thermoplastic. *J. Mater. Sci.* **19**, 1984, pp. 1145-1152

Sato,88	**Sato N., Kurauchi T., Kamigaito O.**, Reinforcing mechanism by small diameter fibre in short fibre composites. *Deformation, Yield and fracture of polymers, 11-14 April, Cambridge, UK* 1988, paper p89.
Sato,91	**Sato N., Kurauchi T., Sato S., Kamigaito O.**, Microfailure behaviour of randomly dispersed short fibre reinforced thermoplastic composites obtained by direct SEM observation. *J. Mater. Sci.* **26**, 1991, pp 3891-3898.
Selvadurai,95	**Selvadurai A.P.S., ten Busschen A.**, Mechanics of the Segmentation of an Embedded Fiber, Part II: Computational Modeling and Comparisons. *J. of Appl. Mech.* **62**, 1995, pp. 98-107.
Stowell,61	**Stowell E.Z., Liu T.S.**, On the mechanical behaviour of fibre reinforced crystalline materials. *J. Mech. Phys. Sol.* **9**, 1961, pp. 242-260.
Suzuki,88	**Suzuki H., Kunio T.**, Influences of fillers, temperature and frequency on the fatigue strength of reinforced nylon 6 and 66. *Proceedings of VI international congress on experimental mechanics and manufacturers' exhibit, 5 - 10 june, Portland, Oregon, USA.* 1988, 902-906.
Toll,93	**Toll S., Anderson P.-O.**, Microstructure of Long- and Short-Fiber Reinforced Injection Molded Polyamide. *Pol. Comp.* **14**, 1993, pp. 116-125.
Tyson,65	**Tyson W.R., Davies G.J.**, A photoelastic study of the shear stresses associated with the transfer of stress during fibre reinforcement. *Brit. J. Appl. Phys.* **16**, 1965, pp. 199-205.
Vincent,86	**Vincent M., Agassant J.F.**, Experimental study and Canculations of Short Glass Fiber Orientation in a Center Gated Molded Disc. *Pol. Comp.* **7**, 1986, pp. 76-83.
Venderbosch,93	**Venderbosch R.W., Nelissen J.G.L., Meijer H.E.H., Lemstra P.J.**, Polymer blends based on epoxy resin and polyphenylene ether as a matrix material for high-performance composites. *Makromol. Chem., Macromol. Symp.* **75**, 1993, pp. 73-84.
Voss,88	**Voss H., Karger-Kocsis J.**, Fatigue crack propagation in glass-fibre and glass-sphere filled PBT composites. *Int. J. of Fat.* **10**, 1988, pp. 3-11.
Wyzgoski,88	**Wyzgoski M.G., Novak G.E.**, A strain energy release rate analysis of the fatigue fracture of nylons. *Pol. prepr., Am. Chem. Soc.* **29**, 1988, pp. 132-133.

References

Wyzgoski,90 — **Wyzgoski M.G., Novak G.E., Simon D.L**, Fatigue fracture of nylon polymers, Part1: Effect of frequency. *J. Mater. Sci.* **25**, 1990, pp. 4501-4510.

Wyzgoski,91 — **Wyzgoski M.G., Novak G.E.**, Fatigue fracture of nylon polymers. Part II, effect of glass-fibre reinforcement. *J. Mater. Sci.* 26. 1991, pp. 6314-6324.

Wyzgoski,92 — **Wyzgoski M.G., Novak G.E.**, Influence of Thickness and Processing History on Fatigue Fracture of Nylon 66. Part I: Crack Propagation Measurements and Part II: Crack tip Morphology. *Pol. Eng. Sci.* **32**, 1992, pp. 1105-1113 and 1114-1125.

Wyzgoski,94 — **M.G. Wyzgoski, G.E. Novak**, Direct measurement of strain energy release rates during fatigue fracture of reinforced nylon 66. *Deformation, Yield and Fracture of polymers, 11-14 April, Cambridge, UK*, 1994, paper 39.

Yu,94 — **Yu Z., Ait-Kadi A., Brisson J.**, Morphological and orientation studies of injection moulded nylon-6,6/Kevlar composites. *Polymer.* **35**, 1994, pp. 1409-1418.

Appendix I, Material description, injection moulding conditions.

1)   Material: Akulon K224-G6
      Supplied by: Akzo-Arnhem / DSM Engineering Plastics, Geleen
      Matrixmaterial: Polyamide 6

Molecular weights:		K123(base matrix material)	K224-G6
	M_n	12900	13200
	M_w	25700	28000
	M_z	39500	44200

Fibre fraction: 30 % Weight

2) Injection moulding conditions

Size:	100x100		90x90		
thickness:	2 mm	5,75mm	2mm	4mm	6mm

Process conditions: not available

Injection temperature, °C:	261		255	255	255
Mould temperature, °C:	40		80	80	80
Injection speed, mm/s:			100	100	120
Injection pressure, MPa:	92,8		84	90	95
Injection time, s:	0,31		0.29	0.49	0.51
Back pressure, MPa:	34,8		35	69	65
Back pressure time, s:	10		10	6	6
Cooling time, s:	15		25	25	25

Appendix II, Example of test sheet.

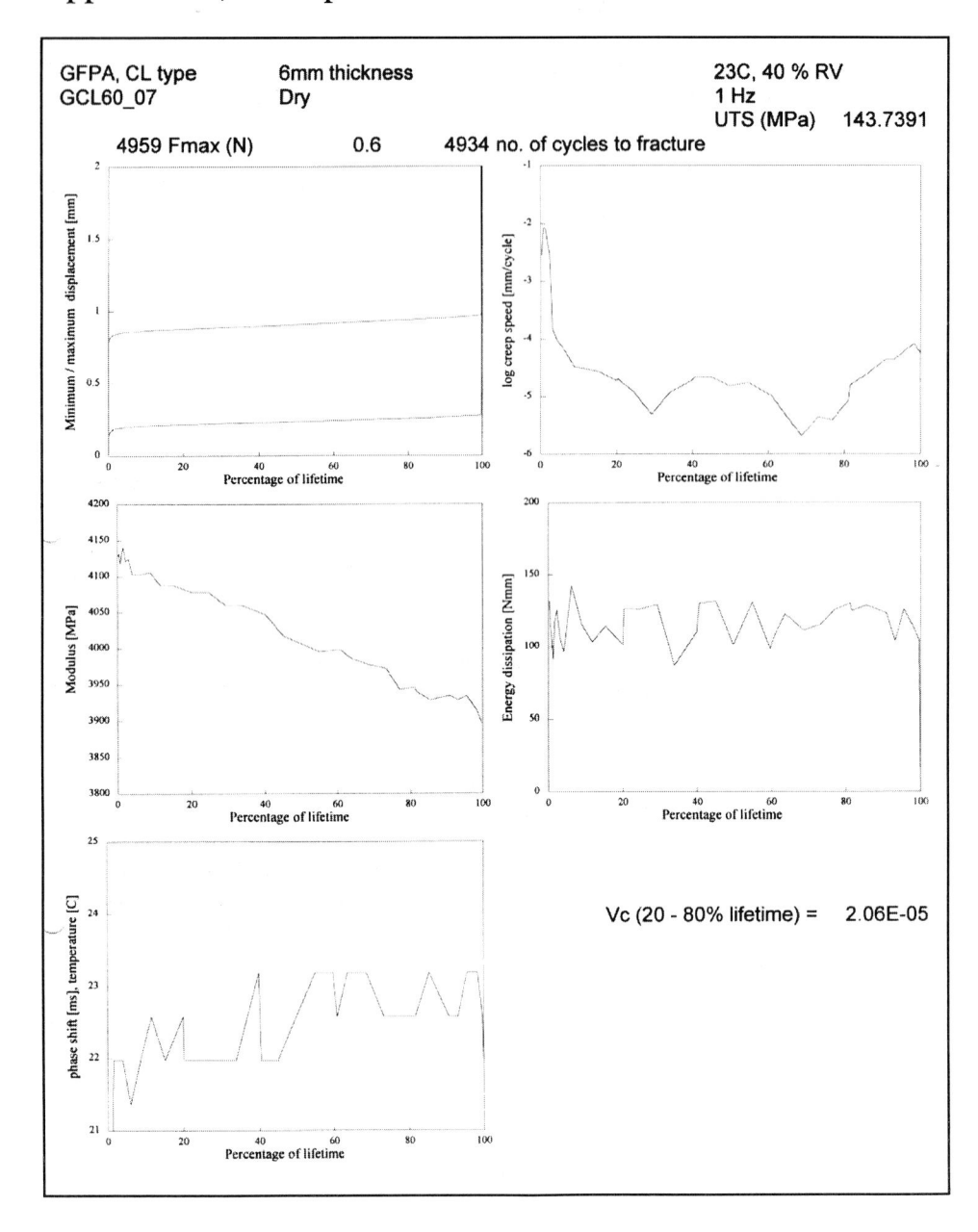

Acknowledgements

The danger with acknowledgements is that one is very likely to forget one of the many people who have been of help. Therefore I would like to begin with thanking everybody who has helped me one way or another, but is not mentioned here!

I would like to express my gratitude towards **Edwin van Hartingsveldt** of DSM Research who was of great help throughout the project. It was also through him that most of the materials, and some of the specimens were obtained. Further cooperation with DSM with especially **Eddy Sham** and **Gertjan de Koning** was very pleasant.

Naturally, thanks must go to **Professor Spoormaker**, who supervised the research and gave me the opportunity to do this work and write the thesis.

Most of the experiments were performed at the Laboratory for Mechanical Reliability (Delft University of Technology), with the technical support of **Wouter Griffioen, Rob Simonis and Duco Smissaert**. Duco was especially helpful in programming a data-retrieval system for the fatigue experiments.

I want to thank also **Rob van den Boogaart** for his support with, amongst other things, the injection moulding of hundreds of plates from which specimens were manufactured.

Frans Oostrum of the aeronautic faculty has been very helpful with the Scanning Electron Microsope of his faculty. It was always possible for me to use this instrument and he helped me many times to use it to the full, so as to make all the fractographs shown.

Research fellow **Nikolay Salienko** from Russia did a project on the micromechanic stresses around the fibre end, which led to an increase in understanding of the failure process.

Many students and trainees, both from the Netherlands and from abroad, have contributed to this work, often working very hard to generate lots of data. They were: **Caroline McCreesh, Cor Visser, Dirk Antonissen, Eugene McGinley, Stefan Fortuin, Gary Kilgore, Manuel Dias, Mathijs Geerling, Peter Nieuwenhuis** and **Theo van Gijssel.**

Not directly contributing to my work, but essential in creating a pleasant atmosphere, were the other Ph.D. students at the Laboratory and the scientific and technical staff, making it possible to discuss various subjects, mostly not at all related to the work here!

Last but not least I want to thank my dear wife **Erika**, who always supported me, especially through the times when my enthusiasm for the project was lowered. Of course the birth of my two children made me see the whole research in a different perspective.

Curriculum Vitae

I was born July 22, 1965 in Winschoten, in the north of the Netherlands. Here I went to school until I was 18, when I got my *Atheneum* diploma in 1983. I went on to study *Werktuigbouwkunde*, mechanical engineering, at the University of Twente, in Enschede. I specialised in Tribology, the science of lubrication, wear, bearing design etc. My training was done at the Koninklijke - Shell Laboratorium Amsterdam (KSLA), where I studied the functioning of a Shell - developed apparatus for measuring the position of the axle in a plain bearing under varying load. This in order to measure the effects of Elasto-viscous additives on film-thickness. I had to test and judge the apparatus, and make proposals for improvements. The graduation study involved the determination of wear of Laser-hardened steel as function of time. I developed a new method to measure wear during the experiment. During my study I worked as a students assistant, instructing students mainly in computer classes.

After my study I went for 5 months to South America, where I visited the most southern city on earth, Ushuaia, visited the dryest desert of the world in Chile, and was on the biggest river of the world, the Amazon.

September 1992 I began my Ph.D. study as an *Assistent in Opleiding* at the Industrial Design Engineering faculty of Delft University of Technology. Here I did my Ph.D. Research on the fatigue of glassfibre reinforced polyamide. During this period I did the post-academic TOP Polymers course of the *stichting Polymeer Technologie Nederland*. Besides the research I had some educational tasks, amongst which coaching of first-year students in design education, giving lectures and coaching of trainees from the Netherlands and from abroad.

In my free time I set up a web-site on my biggest interest, Bugatti, on the rapidly expanding Internet. I also publish an on-line magazine on Bugatti.